Modern Perspectives on Virtual Communications and Social Networking

Jyotsana Thakur
Amity University, India

A volume in the Advances in
Social Networking and Online
Communities (ASNOC) Book Series

Published in the United States of America by
IGI Global
Information Science Reference (an imprint of IGI Global)
701 E. Chocolate Avenue
Hershey PA, USA 17033
Tel: 717-533-8845
Fax: 717-533-8661
E-mail: cust@igi-global.com
Web site: http://www.igi-global.com

Library of Congress Cataloging-in-Publication Data

Names: Thakur, Jyotsana, 1978- editor.
Title: Modern perspectives on virtual communications and social networking /
    Jyotsana Thakur, editor.
Description: Hershey, PA : Information Science Reference, [2018]
Identifiers: LCCN 2017052914l ISBN 9781522557159 (hardcover) l ISBN
    9781522557166 (ebook)
Subjects: LCSH: Internet--Social aspects. l Online social networks. l Virtual
    computer systems.
Classification: LCC HM851 .M634 2018 l DDC 302.30285--dc23 LC record available at https://
lccn.loc.gov/2017052914

This book is published in the IGI Global book series Advances in Social Networking and Online Communities (ASNOC) (ISSN: 2328-1405; eISSN: 2328-1413)

British Cataloguing in Publication Data
A Cataloguing in Publication record for this book is available from the British Library.

All work contributed to this book is new, previously-unpublished material.
The views expressed in this book are those of the authors, but not necessarily of the publisher.

For electronic access to this publication, please contact: eresources@igi-global.com.

Advances in Social Networking and Online Communities (ASNOC) Book Series

ISSN:2328-1405
EISSN:2328-1413

Editor-in-Chief: Hakikur Rahman, BRAC University, Bangladesh

MISSION

The advancements of internet technologies and the creation of various social networks provide a new channel of knowledge development processes that's dependent on social networking and online communities. This emerging concept of social innovation is comprised of ideas and strategies designed to improve society.

The **Advances in Social Networking and Online Communities** book series serves as a forum for scholars and practitioners to present comprehensive research on the social, cultural, organizational, and human issues related to the use of virtual communities and social networking. This series will provide an analytical approach to the holistic and newly emerging concepts of online knowledge communities and social networks.

COVERAGE

- Knowledge Acquisition Systems and Networks
- Challenges of Knowledge Management
- Organizational Knowledge Communication
- Knowledge Communication and the Role of Communities and Social Networks
- Local E-Government Interoperability and Security
- Measuring and Evaluating Knowledge Assets
- Communication and Agent Technology
- Knowledge as a Competitive Force
- Best Practices for Mobile Computing
- Generation of Municipal Services in Multi-Channel Environments

IGI Global is currently accepting manuscripts for publication within this series. To submit a proposal for a volume in this series, please contact our Acquisition Editors at Acquisitions@igi-global.com or visit: http://www.igi-global.com/publish/.

The Advances in Social Networking and Online Communities (ASNOC) Book Series (ISSN 2328-1405) is published by IGI Global, 701 E. Chocolate Avenue, Hershey, PA 17033-1240, USA, www.igi-global.com. This series is composed of titles available for purchase individually; each title is edited to be contextually exclusive from any other title within the series. For pricing and ordering information please visit http://www.igi-global.com/book-series/advances-social-networking-online-communities/37168. Postmaster: Send all address changes to above address. ©© 2019 IGI Global. All rights, including translation in other languages reserved by the publisher. No part of this series may be reproduced or used in any form or by any means – graphics, electronic, or mechanical, including photocopying, recording, taping, or information and retrieval systems – without written permission from the publisher, except for non commercial, educational use, including classroom teaching purposes. The views expressed in this series are those of the authors, but not necessarily of IGI Global.

Titles in this Series

For a list of additional titles in this series, please visit:
https://www.igi-global.com/book-series/advances-social-networking-online-communities/37168

Exploring the Role of Social Media in Transnational Advocacy
Floribert Patrick C. Endong (University of Calabar, Nigeria)
Information Science Reference • ©2018 • 307pp • H/C (ISBN: 9781522528548) • US $195.00

Online Communities as Agents of Change and SocialMovements
Steven Gordon (Babson College, USA)
Information Science Reference • ©2017 • 338pp • H/C (ISBN: 9781522524953) • US $195.00

Social Media Performance Evaluation and Success Measurements
Michael A. Brown Sr. (Florida International University, USA)
Information Science Reference • ©2017 • 294pp • H/C (ISBN: 9781522519638) • US $185.00

Political Scandal, Corruption, and Legitimacy in the Age of Social Media
Kamil Demirhan (Bülent Ecevit University, Turkey) and Derya Çakır-Demirhan (Bülent
Ecevit University, Turkey)
Information Science Reference • ©2017 • 295pp • H/C (ISBN: 9781522520191) • US $190.00

Power, Surveillance, and Culture in YouTube™'s Digital Sphere
Matthew Crick (William Paterson University, USA)
Information Science Reference • ©2016 • 317pp • H/C (ISBN: 9781466698550) • US $185.00

Social Media and the Transformation of Interaction in Society
John P. Sahlin (The George Washington University, USA)
Information Science Reference • ©2015 • 300pp • H/C (ISBN: 9781466685567) • US $200.00

New Opportunities for Artistic Practice in Virtual Worlds
Denise Doyle (University of Wolverhampton, UK)
Information Science Reference • ©2015 • 322pp • H/C (ISBN: 9781466683846) • US $185.00

For an entire list of titles in this series, please visit:
https://www.igi-global.com/book-series/advances-social-networking-online-communities/37168

701 East Chocolate Avenue, Hershey, PA 17033, USA
Tel: 717-533-8845 x100 • Fax: 717-533-8661
E-Mail: cust@igi-global.com • www.igi-global.com

Table of Contents

Detailed Table of Contents

Chapter 1
Zoheir Ezziane, Al Ain Women's College, UAE

The aim of this chapter is to illustrate how social networking could be used as a tool to empower people and organizations to get connected and share similar ideas and endeavors. It demonstrates the benefits when organizations employ social networking as an e-service tool to engage both consumers and businesses alike. In this chapter, a special focus is attributed to Al Ain Distribution Company (AADC), a wholly owned subsidiary of Abu Dhabi Water and Electricity Authority, in the United Arab Emirates (UAE). AADC has implemented novel e-services for the purpose of improving customer services and incorporate social networking within its existing management information system (MIS). This work has been instrumental in not only showing advantages of using social networking at AADC but also helping the company to address various consumer needs and enhancing its e-services.

Chapter 2
Sarah Bouraga, University of Namur, Belgium
Ivan Jureta, Fonds de la Recherche Scientifique, Belgium & University
of Namur, Belgium
Stéphane Faulkner, University of Namur, Belgium

Online social networks (OSNs) such as Facebook and LinkedIn are now widely used. They count users in the hundreds of millions. This chapter surveys popular social networks in order to present a pattern of recurring functional requirements as well as non-functional requirements, and a model of that pattern in the i* requirements modelling language. The pattern can serve as a starting point for requirements engineering of new OSNs. The authors test their model by applying it to a popular OSN, namely Twitter.

Chapter 3

Elspeth McKay, RMIT University, Australia
Allaa Barefah, RMIT University, Australia
Marlina Mohamad, Universiti Tun Hussein Onn Malaysia, Malaysia
Mahmoud N. Bakkar, Holmes Institute, Australia

Many people are finding it relatively easy to engage with courseware development and simply upload it to the internet. The trouble with this approach is there are no quality controls to ensure the impending instructional strategies are designed well. This chapter presents a set of research projects that incrementally focus on instructional strategies as they apply in today's information communications technology (ICT) tools. They commence with a simple investigation into online courseware matching cognitive preferences. Next, a project extends this principle and takes the research to a government-training context to concentrate on dealing with a broader range of stakeholders to customize training for a user-centered environment. Then the authors present a project designed for mobile healthcare training on an iPad. The final research project synthesizes the current instructional systems design (ISD) thinking to promote a prescriptive information systems (IS)-design model for educational courseware, offering it to the ISD research community as an extension to the ADDIE training development model.

Chapter 4

Azza Abdel-Azim Ahmed, Abu Dhabi University, UAE & Cairo
University, Egypt

This research aimed to explore types of online social capital (bridging and bonding) that the Emiratis perceive in the context of social networking site (SNS) usage. A sample of 230 Emiratis from two Emirates, Abu Dhabi and Dubai, was used to investigate the hypothesis. The results showed that WhatsApp was the most frequent SNS used by the respondents. Also, a significant correlation of the intensity of social networking usage and bridging social capital was found, while there was no significant association between SNS usage and bonding social capital. The factors determined the SNSs usage motivations among the respondents were exchange of information, sociability, accessibility, and connections with overseas friends and families. Males were more likely than females to connect with Arab (non-Emiratis) and online bonding social capital. Both genders were the same in their SNSs motivations and online bridging social capital.

Chapter 5

Sarah Bouraga, University of Namur, Belgium
*Ivan Jureta, Fonds de la Recherche Scientifique, Belgium & University
of Namur, Belgium*
Stéphane Faulkner, University of Namur, Belgium

The last decade has seen an increasing number of online social network (OSN) users. As they grew more and more popular over the years, OSNs became also more and more profitable. Indeed, users share a considerable amount of personal information on these sites, both intentionally and unintentionally. And thanks to this enormous user base, social networks are able to generate recommendations, attract numerous advertisers, and sell data to companies. This situation has sparked a lot of interest in the research community. Indeed, users grow more uncomfortable with the idea that they do not have full control over their own data. The lack of control can even be amplified when a user holds an account on various OSNs. The data she shares is then spread over multiple platforms. This chapter addresses the notion of portable profile, which could help users to gain more control or more awareness of the data collected about her. In this chapter, the authors discuss the advantages and drawbacks of a portable profile. Secondly, they propose a conceptual model for the data in this unified profile.

Chapter 6

*Shamsher Singh, Banarsidas Chandiwala Institute of Professional
Studies, India*
*Deepali Saluja, Banarsidas Chandiwala Institute of Professional
Studies, India*

In the information age, social media is growing rapidly and at a faster pace. Social media is playing an important role in the day-to-day life of individuals. Using social media has become the everyday routine. Many social media sites display different types of advertisements by which the decision-making process is generally getting affected. Social media is much more than just a medium of sharing information. The present study is an attempt to understand how social media affects the decision-making process of consumers and the impacts of various marketing strategies used by firms on social media. The study employs the survey method to collect primary data from 200 customers who have been regularly using social media. Factor analysis and ANOVA has been used to gain insights in the study. The selected respondents are assumed to represent the population in the urban areas of Delhi.

Chapter 7

Interactive digital storytelling (IDS) allows a human user to become an active part in a story and to affect how the story unfolds. To understand IDS systems, we need to consider the partakers present in them as well as their roles and interconnections. In this chapter, the authors discern four partaking entities—interactor, author, developer, and storyworld—and describe both their affiliated sub-entities as well as their relationship to one another. Based on both reviewing relevant literature and analyzing existing IDS systems, the ontology presented here provides a cohesive view into the current state of both theoretical and practical research.

Chapter 8

This chapter examines the social media content posted by a woman Indian chief executive officer (CEO) on Twitter. The active involvement of CEO in communication activities influences the business effectiveness, performance, and standing of the business headed by her. Rstudio and Nvivo, two analytical tools, were used for different analysis such as tweets extraction and content analysis. The findings show the various themes in CEO communication which are categorized in different sectors in terms of her personal views (feelings and status updates), political views, and social concerns (ranging from education, women empowerment, governance, and policy support). The chapter extends the theoretical and empirical arguments for the importance of CEOs' social media communications. Finally, this research suggests that with a well-planned and strategic social media use, CEOs can create value for themselves and their businesses.

Preface

Modern Perspectives on Virtual Communications and Social Networking is a humble narration of my encounter with thrill, enthusiasm, joy and harmony of infinite reservoir of the wisdom that becomes comprehendible on the strength of the power of mind and the grace of social networking. In life, being persistently agog with a strong impulse to improvise shall usher one on to the path of success and glory. The book is an avid plan of modern perspectives on internet, social networking and virtual communities to walk on trails of excellence. The narration has a message universal in its appeal and purpose. It formulates us to practice innovation as a pragmatism to self and humanity.

The rubrics are shifting and so is the fact. The society who are privileged enough to have perceived the 'controlled' media and the embryonic new 'uncontrolled' media market would endorse for the penetration and scale of the transformation that media markets have come across. The deteriorating antagonism and market disintegration has left the researchers looking for the mitigation. The whole of the media attentions, precisions and rigors have trickled down to conceiving techniques and intends to generate and retain the customer. If corporates could get a magic wand, then the wish would be to latch the customers with its brands firmly and forever. The current state of the society does not run on social pressures but candor and self-possession govern the society of relationships now. This is applicable to all business relationships too. Accordingly, the hunt for customer base is leading the marketers and businesses to take shelter in the power of new media i.e. the Internet. The magic wand of businesses in this century is none other than Social Networking, which can capitalize any business while making powerful relationships with the customers. Virtual Communities and Social Networking can provide an effective escape route from the limits imposed by the traditional media and spread the idea to the greater scope of value. Communities and relationships are gaining momentum through social networking sites. Communities with enviable social triumphs imprint images in attitudes and minds of the public of similar viewpoints. These images are the factors that spell trust, commitment and transparency. Social networking ultimately latches customers into a relationship by creating virtual bonds. Therefore the objective of writing this book is to gather latest views and thoughts on virtual communities and social networking to gain knowledge about how the world is changing its horizons.

Social Networking means communicating and sharing your ideas with individuals and groups at the same time. To understand the meaning of communicating successfully with individuals and groups on social networking sites, we must develop both a foundation of communication skills and an understanding of key elements critical to achieve success on such platforms. This book is to provide a structure for learning these necessary expertise in a way that emphasizes the exclusivity of each platform and each individual within the group. Successful social networking starts with strong relationships. This work highlights the crucial aptitudes of social networking in constructing and retaining these relationships. In order to unite all followers living in virtual communities and take decisions or building an opinion or image is the spirit of social networking. Looking at the dynamics of Social Networking, it is easy to say why it can be a challenging phenomenon to understand. Online Social Networks facilitates freedom of expressing views and sharing ideas. It is for the provider to understand the conceptual framework and design the apt platform necessary for sharing and implementing successful communication. This requires many tangible and intangible factors to be keep in mind to meet all challenges.

I had the following main goals in compiling this book:

- **Precision:** This book is the result of the efforts of many authors who were facilitating and researching online social networks. It is important to understand skills that are based on precision in the field of content and design of social networks. This book places a clear emphasis on innovative skills first but also ensures that those skills are based on rigorous and current research.
- **Sharing of Experiences:** To describe and explain social networking communication concepts, this book uses real stories to explain to get insights from various platforms. These stories will help distinguish between effective and ineffective communication practices and to choose one as well as help to identify the social capital to adopt to improve their online social networks.
- **Designing Tactic:** Most importantly the designing of the particular platform for sharing content in terms of elements that can be used to evaluate the effectiveness of the platform used. These elements—compatibility with the devices, links with other websites, individual uniqueness of the platform, structure of the platform and being user friendly are included in the present work.

- **Sociability:** This book explores most exciting feature of online social networks. It provides users to connect with their friends and relatives. This is the reason behind the popularity of online social networks at large level. It is bringing the world community closer. Also revealing culture, traditions, discoveries and many events to the virtual communities. This is an instrument which is uniting people of all communities of similar ideas on one platform.

- **Accessibility:** The easy access of social networking platforms is another reason to make them usable for many purposes such as information exchange e.g. news, advertising, promotion of ideas and services, trade & businesses, and research & development. Use of online social networks for all the above purposes requires easy access to internet which is now available to everyone.

- **Planned Communication Strategies:** This book is dedicated to find out the requirements to develop a strong platform where people can share their ideas and viewpoints. To construct such kind of platform one must understand consumer needs and behavior. This data is required to develop planned social media communication strategies.

APPROACH

The title of this book, *Modern Perspectives on Virtual Communications and Social Networking*, expresses many central constituents of virtual communities and social networking: creation of content, construction of user friendly platforms, use of online platform for everyday businesses and evaluation of effectiveness etc. This book will determine the exclusive dynamics of online social networks, the social support people need to be successful, the role of virtual communities to share ideas on multifaceted platforms, and freedom to express views and encourage people towards development. By evaluating each given site and the viewpoints, readers come to know the vigorous capacity of each one and learn to consider each one as a unique communication opportunity. This book underlines to consider each condition as exceptional, appraise the unique and apt technique and procedure to solve, and implement that efficiently. Basically the approach of compiling this book is to provide a solution for the readers who are inquisitive to know the essence of virtual communities and role of social networking in everyday life of a human who is a social animal.

ORGANIZATION

The book covers the exciting horizons of science and technology, innovative models and designs, introduces to the adventure of building institutional and social excellence along with highly revealing stories of the society. Till recently, communication has been understood to convey messages that could add a benefit to the society. Now with the advent of internet virtual communities and social networking has become the tool to empower people and organizations to get connected and share similar ideas and endeavors. The last few decades have seen an increasing number of Online Social Network (OSN) users. As the popularity is growing over the years, OSNs became also more and more money-spinning. For such a growing consumer base there should be well designed models to make OSNs user friendly and more interesting for users.

This book covers the major areas of Virtual Communities and Social Networking. This book is organized into eight chapters. A detailed description of each chapter is as follows:

Chapter 1 titled 'Social Networking: A Tool for enhancing E-Services' is a true meaning of social networking which says that it is a tool for improving e-services. It illustrates how social networking is used as a tool to empower people and organizations to get connected and share similar ideas and endeavors. It portrays the benefits when organizations employ social networking as an e-service enhancement tool to engage both consumers and businesses alike. In this paper, a special focus is attributed to Al Ain Distribution Company (AADC), a wholly owned subsidiary of Abu Dhabi Water and Electricity Authority, in the United Arab Emirates (UAE). AADC has implemented novel e-services for the purpose of improving customer services and incorporate social networking within its existing Management Information System (MIS). This chapter has been instrumental in not only showing advantages of using social networking at AADC but also helping the company to address various consumer needs such as to add value to work environment such as improving efficiency, and employee work engagement. Furthermore, time management, meeting experts, discussing professional opinions and recommending the right training are some of many values recognized and could be helping the productivity of the company at the end of the stream. This chapter explains that Social Media is a nonstop vigor of creation, it creates competitive environment among employees to encourage them to research and invent more which is the most satisfactory source of happiness. It illustrates that over 60% of staff feels it would enhance their experience if they had awareness sessions showing values and benefits in global manner and how this technology is going to serve and fulfill company needs and help employees accomplish their work efficiently and effectively and enhancing its e-services.

Chapter 2 titled 'Functional and Non-Functional Requirements Modeling for the Design of New Online Social Networks' is an instrumental work included in this book. This chapter explores the basic building requirements for Online Social Networks (OSNs), such as Facebook and LinkedIn, which are now widely used. This paper surveys popular social networks in order to present a pattern of recurring functional requirements as well as non-functional requirements, and a model of that pattern in the requirements modelling language. The pattern can serve as a starting point for requirements engineering of new OSNs. The authors tested their model by applying it to a popular OSN, namely Twitter. In this chapter, the authors offered an analysis of recurrent features in relation to several OSNs, namely Facebook, Flickr, LinkedIn, MySpace, Pinterest, Tumblr, Twitter and YouTube. They also represented the pattern of features using an existing requirements engineering language, namely i*. This is the first research to address OSNs in terms of (1) identifying requirements implemented by OSN features, and (2) modeling these requirements using i*. And secondly, it can have implications for the design of future OSNs; more specifically for the RE phase.

Chapter 3 titled 'Advances in E-Pedagogy for Online Instruction: Proven Learning Analytics Rasch Item Response Theory' is an important piece of knowledge added to the book. This chapter explains that there are no quality controls to ensure that to develop courses under information communication technology (ICT), appropriate instructional strategies are not available. The authors presented a set of research projects that focus on instructional strategies (ePedagogy) through this chapter. They investigated into online courseware matching cognitive preferences. Next, a project extends this principle and takes the research to a government-training context to concentrate on dealing with a broader range of stakeholders to customize training for a user-centered environment. Then they presented a project designed for mobile healthcare training on an iPad. Their final research project synthesizes the current instructional systems design (ISD) thinking to promote a prescriptive information systems (IS)-design model for educational courseware, offering it to the ISD research community as an extension to the ADDIE training development model. According to the authors there can be no doubt that ICT has brought us a tremendous number of ways to deliver our courseware. Along with this feast of multi-media delights is the relative ease which many people experience these days putting their learning/ training programs online. The authors of the paper stressed upon the key points that the researchers should initiate to bring new ISD models according to the need to reflect the way people want to receive their information these days.

Chapter 4 titled 'Online Social Capital Among Social Networking Sites' Users' aimed to explore types of online social capital (bridging and bonding) that the Emiratis perceive in the context of social networking site (SNS) usage. The author used a snow-ball sample of 230 Emiratis from two Emirates, Abu Dhabi and Dubai. The results showed that WhatsApp was the most frequent SNS used by the respondents. Also, a significant correlation of the intensity of social networking usage and bridging social capital was found, while there was no significant association between SNS usage and bonding social capital. This chapter gave the insights that both genders were found same in their SNSs motivations and online bridging social capital. In this chapter it seems that the SNS usage habits have changed rapidly throughout the past few years. The chapter suggests that the type of relationships within the social network can predict different kinds of social capital. The results showed that the Internet is particularly useful for keeping contact among friends who are socially and geographically dispersed. It predicts that communication is lower with distant than nearby friends. Also, the research revealed that the more diverse the social categories of "relationships" and "nationalities", the more bridging social capital can be predicted. Bonding social capital is associated positively with "relationships", but negatively with "nationalities". This means that Emiratis are keen to use SNS to broaden their social networks rather than get/provide emotional support. The results of this chapter showed an association between intensity of SNS connection and bridging social capital, while no significant correlation was found between intensity of SNS connection and the bonding social capital. It also indicates that bridging social capital was the most valued use of Facebook. It was also explained that the intensity of Facebook usage can help students accumulate and maintain bridging social capital. This form of social capital; which is closely linked to the notion of "weak ties", seems well suited to social software applications, because it enables users to maintain such ties cheaply and easily. The Facebook intensity is positively correlated with all types of social capital. In the current study, four factors were found to determine the respondents' motivations of SNS usage. These factors are: Exchange information, Sociability, Accessibility (which refers to easy way to access new information), and Connecting with overseas friends and families.

In addition, the findings of the current research revealed that the four SNS motivation factors were associated with increase online bridging and bonding. Moreover, they provided evidence of the sociability motivation mediating the relationship between Facebook habits and online bonding and bridging.

Chapter 5 titled 'Users Holding Accounts on Multiple Online Social Networks: An Extended Conceptual Model of the Portable User Profile' is an important chapter included in this book. It explores that a few last decades have seen an increasing number of Online Social Network (OSN) users. As they grew more and more popular over the years, OSNs became also more and more profitable. Users share a considerable amount of personal information on these sites, both intentionally and unintentionally which is adding to this enormous user base, social networks are able to generate recommendations; attract numerous advertisers; and sell data to companies. This situation has sparked a lot of interest in the research community. The lack of control can even be amplified when a user holds an account on various OSNs. The data user shares is then spread over multiple platforms. This paper addresses the notion of portable profile, which could help users to gain more control or more awareness of the data collected about them. In this chapter the authors discussed the advantages and drawbacks of a portable profile and they proposed the "Portable User Profile" (PUP) conceptual model. This model is meant to gather all the data/information a user shares on every OSN he/she uses. The PUP should be accessible by both the user, and the OSN. Before introducing the conceptual model, the authors discussed the advantages and limitations of such a profile. An integrated profile is richer in terms of information than a single OSN account, and this enriched profile probably leads to better, more accurate recommendations. Also, the users could have access to all the information the OSNs gather about them; and thus the users could have more control over their own data. Nevertheless, with an integrated profile the user cannot choose an OSN where they would be more active, where they would share more information. The proposed PUP conceptual model is composed of two main classes: Explicit and Implicit data. Each of these classes can be specialized into two classes: Profile and Relationship data; and Posts and Activity data respectively. Those four classes are specialized into several other classes. They represented this conceptual model as a UML Class diagram.

Chapter 6 titled 'Effects of Social Media Marketing Strategies on Consumers Behavior' explores that social media is growing rapidly and at a faster pace. Social media is playing an important role in the day to day life of individuals. Many social media sites display different type of advertisement by which decision making process is generally getting affected. The present study is an attempt to understand how the social media affect the decision making process of consumers and impact of various marketing strategies used by firms on social media. The authors of this study employ the surveys methods to collect primary data from 200 customers who have been regularly using social media. Factor Analysis and ANOVA has been used for having insights in the study. The selected respondents are assumed to represent the population in the urban areas of Delhi. The empirical evidence based on ANOVA indicates that there are not significant variance in customer responses based on the gender, occupation and income. This indicated that the customer irrespective of their gender, occupation and income view the social media networking site in similar way. However there are significant variance on the bases of age and education of the respondents. This study bring out the significant characteristic of social media " such as time spent on site, type of advertisement viewed, specific reason for using social media website and impact on the professional life". It was found in this chapter that brand/ organization which are present on social media website does influences purchase decision making of customers, which suggests that organization should carefully choose the contents of the advertisement on social media. Factor analysis have brought six factor representing different elements of social networking sites These factors are "information factor, user factor, preference factor, impact factor, decision factor and advertisement factor". This brings out the important factor regarding how the manager should plan their marketing strategy depending upon the focus of the strategy.

Chapter 7 titled 'The Digital Campfire: An Ontology of Interactive Digital Storytelling' discovers that Interactive digital storytelling (IDS) allows a human user to become an active part in a story and to affect how the story unfolds. To understand IDS systems the authors need to consider the partakers present in them as well as their roles and interconnections. In this chapter, the authors discern four partaking entities – interactor, author, developer, and storyworld – and describe both their affiliated sub-entities as well as their relationship to one another. The ontology presented here provides a cohesive view into the current state of both theoretical and practical research. In the era of digital culture, it has become more important to recognize the sources of interaction and see how they affecting all the actors in the presented ontology. Instead of being just a passive audience, the interactors have an active role in creating the story. Also, the storyworld includes many computer-controlled mechanisms – most notable intelligent characters – that provide their input into the mixture. As in any classification, drawing the lines is not always easy or unambiguous. For instance, the border between the roles of developer and the author can be very vague. When an IDS system is created by a single individual, this person is then, naturally, both the author and the developer. In larger organizations, people have mixed roles: some may be exclusively developers, creating the infrastructure only, and some may be purely authors focusing solely on the story. Another problematic area can be the border between the author and the interactor. For instance, in tabletop roleplaying games the author is the game master who has created the adventure and runs the game for a group of people – but the game master also experiences the story together with the players.

IDS is in a constant flux, because the technology is still far from mature. Nevertheless, we are seeing how the theoretical framework and understanding is getting firmer and more cohesive. Also, the IDS systems – albeit often developed for quite limited setups – are propelling the field forward. The most likely utilizer of this work will emerge from video games, which have, for a long time, shown a keen interest in including IDS in the game design. That is not to say that IDS will be used only for entertainment purposes, but it will show its potential in more serious applications ranging from physical and psychological well-being to teaching skills, sharing culture and increasing the understanding of the human condition. Human beings crave for stories. That is why we want to tell them and hear them, affect them and be affected by them. IDS is the next step where the evolution of digital media is leading to – an era of a digital campfire.

Chapter 8 titled 'Tweeting About Business and Society: A Case Study of an Indian Woman CEO' examines the social media content posted by a woman Indian chief executive officer (CEO) on Twitter. The active involvement of CEO in communication activities influences the business effectiveness, performance, and standing of the business headed by her. The authors used Rstudio and Nvivo, two analytical tools for different analysis such as tweets extraction and content analysis. The findings show the various themes in CEO's communication which are categorized in different sectors in terms of her personal views (feelings and status updates), political views and social concerns (ranging from education, women empowerment, governance, and policy support). The paper extends the theoretical and empirical arguments for the importance of CEOs' social media communications. Finally, this research suggests that with a well-planned and strategic social media use, CEOs can create value for themselves and their businesses. This study shows the various different perspective of discussion and opinions of an Indian woman CEO in her communication on Twitter. Interestingly, CEOs are more likely to share the newspaper article than the long blog post. As findings show the presence of broad different categories in Ms. Shaw's tweets, it reflects the balanced approach to creating content and engaging with public on social media. Her content is quite vocal about the social issues affecting the local community and even at the national level. She follows the accountability of the public relations professional to connect with different stakeholders (public and media) in relationship building activities. Her views also reflect lobbying for better support in research and development. This broader interaction strategy by a CEO may work better if it accompanies a suitable and strategic social media communication way involving severe ethical issues, and moral challenges.

Jyosana Thakur
Amity University, India

Acknowledgment

In an endeavor like this, the entire credit for the work goes to the authors. I wish to congratulate all the authors who have contributed in this outcome. I wish to thank Zoheir Ezziane, Sarah Bouraga, Ivan Jureta, Stéphane Faulkner, Elspeth McKay, Allaa Barefah, Marlina Mohamad, Mahmoud N. Bakkar, Azza Abdel-Azim Ahmed, Shamsher Singh, Deepali Saluja, Jouni Smed, Tomi "bgt" Suovuo, Natasha Trygg, Petter Skult, Harri Hakonen, Ashish Kumar Rathore, Nikhil Tuli, and P. Vigneswara Ilavarasan. I would like to acknowledge and appreciate the support provided by Maria Rohde at every step.

Jyosana Thakur
Amity University, India
June 2018

Chapter 1
Social Networking:
A Tool for Enhancing E-Services

Zoheir Ezziane
Al Ain Women's College, UAE

ABSTRACT

The aim of this chapter is to illustrate how social networking could be used as a tool to empower people and organizations to get connected and share similar ideas and endeavors. It demonstrates the benefits when organizations employ social networking as an e-service tool to engage both consumers and businesses alike. In this chapter, a special focus is attributed to Al Ain Distribution Company (AADC), a wholly owned subsidiary of Abu Dhabi Water and Electricity Authority, in the United Arab Emirates (UAE). AADC has implemented novel e-services for the purpose of improving customer services and incorporate social networking within its existing management information system (MIS). This work has been instrumental in not only showing advantages of using social networking at AADC but also helping the company to address various consumer needs and enhancing its e-services.

DOI: 10.4018/978-1-5225-5715-9.ch001

INTRODUCTION

With the increasing advancements in modes of communication, the world has become small. In the era of internet, fast messaging and social networking sites, the communication and interactions without Facebook, Twitter, and Myspace cannot be even imagined. The social media (SM) facilitates people needs within a single space, provides platform to share their views, ideas, creating their own content and many more things to enjoy and communicate in effective manner (Asur and Huberman, 2010; Ezziane and Al Kaabi, 2015; Stark and Krosnick, 2017). From organizational aspect, companies with SM are getting spread efficiently over thousands of different customer's layers from different locations in the world. Team performance is also being measure through the use of social networking, and its positive relationship with leadership (Mukherjee, 2016). With help of SM, no matter where your company location is and how much capabilities in the market it has, small business still can compete with giants in the market and sometimes they tend to have same number of fans than big companies (Hanafizadeh et al., 2012).

Abu-Dhabi Distribution Company (AADC) is interested in supporting their management information systems (MISs) to integrate social networking into its organizational business activities to develop new IT business processes and improve their online ecommerce service capabilities. In order to achieve the maximum results, AADC management decide to conduct a pilot study to see the impact of having social networking platforms used within AADC environment. One of the results which AADC hope to prove in this research, increasing a wider network of both business contacts and relationships with companies and consumers of common interests and activities. AADC research team is going to see how helpful to use social networks (SNs) like Facebook, Twitter, LinkedIn and Google+ to share information with customers and learn about consumer complaints about products and services that require improvement.

AADC was founded in 1995 by the Abu Dhabi government and in 2008 joined the Abu Dhabi Sustainability Group (ADSG) to help provide clean water and electricity to the entire population of the city with 1.5 million people. AADC uses desalinated seawater as the drinking water for supplying all Abu Dhabi customers and the chemically treated wastewater is reused for irrigation and landscaping purposes. AADC continuously researches innovative methods of better satisfying consumers by developing techniques that reflect global environmentally-friendly policies. It also crates new ways to efficiently distribute and produce electricity and water to its customers. In addition, it is currently focused on upgrading all aspects of their customer services to find out ways they can better serve the public. AADC is trying to be more convenient with online processes related to paying monthly utility bills and inform consumers of ways to save energy and water (AADC, 2013).

Some of the main problems of AADC is encountering issues when incorporating SM platforms into their MIS processes to improve productivity and efficiency of AADC products and services. AADC believes that as much there is a wider network of customers, having database is going to be richer in matter of customer profiling, feedback, and complaints and therefore there will be high chances of enhancing products and services. In other hand, using SM as part of employee daily duty will require a set of skills to manipulate search engines. Many UAE national employees are very skilled at social networking for their personal lives and "Facebook" found to be the most favored SN in UAE (Al-Jenaibi, 2011). The use of SM and personal networks are also on the rise in the Arab world (Alreyaee and Ahmed, 2015; Van Tubergen et al., 2016).

LITERATURE REVIEW

SNs are online platforms over the Internet and World Wide Web that provide services for global communications among people who share common activities or interests. The development of social networking as an innovative method of meeting new people, building business or social contacts and to attract others to participate in various activities or events has become an effective marketing technique worldwide (Hanafizadeh et al., 2012; Miranda et al., 2014).

According to Haythornthwaite (2005), the history of SNs began with people online using various Internet forums to communicate with others within their existing social circle. SN sites evolved later into online forums like singles sites for networking people who did not previously know each other who either shared common interests or wanted to meet someone to date. The first official social networking site was created in 1978 by Suess and Christensen and called the Computerized Bulletin Board System (CBBS), which was more of an online forum for private users only. However, it was many years later that the SN systems of today were introduced to the public. When Internet Protocol Systems (IPSs) allowed for mass global online usage, social networking exploded into a major online phenomenon (Boyd & Ellison, 2007).

By the 1980s, many virtual online forums like the Beverly Hills Internet site (which was later renamed GeoCities) were launched with special chat rooms, profiles, member pages and photo galleries. In 2003, Jonathan Abrams started Friendster as the first innovator of social networking sites, which had over three million members within six months, paving the way for Myspace and Facebook. In 2003, 250 Friendster members got 10 friends each to join them to start up Myspace, which became the first really famous social networking site for two years. However, by 2004, when Mark Zuckerberg launched the Facebook SM website from Harvard

University campus by hacking into the college student profile pages, he developed a unique way to link people and allow for comments, opinions and postings that others could read online.

All previous social networking websites could not compare to the huge volume of members Facebook attracted with this inventive technique or viewing people for either personal or social relationships within a few years. Facebook continued to develop its overall capabilities to add many new features that allowed for photo sharing, tagging pictures and sending them to others, requesting friends and linking to their SNs to continuously add new contacts.

There are several advantages for using SNs that improve efficiencies, develop core competencies and increase employee engagement in the workplace. Many companies now realize how SNs can improve employee morale by making their work more interesting and giving them a global perspective of consumer demands. Some of the main advantages of using SNs include (Dyer, 2013):

- Meeting people from all over the world for social and business contacts
- Learning new information and knowledge related to common interests and activities
- Capacity and forum to post personal or business ideas and opinions on a daily basis that will generate global discussions and debates with useful knowledge that will provide valuable insight into various important issues
- Continuous new links to contacts with similar links and their comments, blogs, feedback and networks

There are many different types of special purpose SNs that serve a particular market that companies can utilize to maximize their online presence. The majority of business uses of online SNs include increasing awareness of company brand name, products and services, e-commerce shopping, e-banking or billpay, virtual marketplaces, and for advertisements that links to company websites providing information on products and services (Khanna & Wahi, 2014). Other corporate uses for SM include researching global markets, facilitating mergers and acquisitions, controlling data information and developing online databases (Hanafizadeh et al., 2012).

The future of SNs includes sites like Facebook wanting to open up their sites to allow developers to use their social database to create ways for users to keep their identities as they cross over different Facebook platform applications (Wikle & Comer, 2012). There are many problems with privacy issues and complex IT architectural platform problems, however, the innovators of Facebook and other sites are certain it is possible. The global laws and regulations have to be stricter to prevent hackers from stealing online data to protect users worldwide.

Management Information System (MIS) and AADC's MIS Strategies

In today's organizations, the information systems (IS) departments are managed as a part of the IT department providing interrelated computer services. IS relates to complementary computer and IT software and hardware networks companies utilize for gathering, filtering, creating, distributing and processing data. IS uses theoretical frameworks related to computation and information to develop computer science and business processes. The study of IS refers to development of useful business models and frameworks with associated algorithm processes under the disciplines of computer science. IS focuses on supporting management, decision-making and business operations so people can interact more effectively with various types of technologies to support different business processes (McNurlin, 2009).

MIS supports Information and Communication Technology (ICT) approaches for companies that want to integrate their social networking strategies to monitor consumer feedback and link the information to their future strategic planning. This assists management in foreseeing the possible weaknesses and difficulties they must address and resolve in order to improve customer services and organizational efficiencies. Many global governments have been supporting e-government ICT planning using MIS to integrate social networking into its organizational business activities to develop new IT business processes and improve their online ecommerce service capabilities (Ezziane & Al Shamisi, 2013; McKeen, 2011).

AADC's MIS departmental structure involves the management of the roles, divisions, processes, and governance of IS, IT and ICT systems. The relationship between AADC's governance structure and IS governance structure is that they both have a centralized hierarchical structure. AADC IS Department has a structure as a separate department under central control. The structure of the IS Department in AADC is a hierarchy of top management, assistant manager, MIS IT specialist, and 10 employees having an IT Bachelor's degree from universities around the world. This structure is the best fit for the organization because it helps people know who they must go to if they have questions about their responsibilities, what the management expects from them, and what they need to do in order to prepare for getting promotions in the future. This structure also helps employees know what position they could be working toward above their current job in order to gain the knowledge and job skills needed (Salem, 2009).

AADC's main business functions focus on service provisions related to documentation processing and bill payment. They provide various online e-commerce customer services, including governmental bill payment options for handling home and business applications and for paying monthly residential and commercial utility bills. AADC's business involves many functional and cross functional processes and services for the public to increase convenience for them and reduce waiting time for processing paperwork. Their governance structure includes local private and public sector partnerships and projects related to global best practice benchmarks, and international corporations for online financial transaction services and products.

MIS Efficiencies Using SNs and Supporting E-Government ICT

AADC's MIS has aided in developing their business functions and offerings with advanced e-government options so consumers have faster and easier access to online services, government information, discussion forums and departmental strategies for the future. AADC's MIS strategies facilitate public processes and procedures for all divisions to increase efficiency and productivity and reduce overhead costs. The MIS also improves interrelations and communication between government departments to create better policymaking for upgrading reforms. AADC's MIS allows for more direct communication between them and the public so consumers can be more a part of the overall decision-making processes.

AADC's MIS helps reduce consumer documentation processing time for applications and bill payments. MIS allows for more ecommerce possibilities for all elements of the e-government portals and decreases consumer costs by increasing automation and reducing government transaction and processing costs (Dahi and Ezziane, 2015; Salem, 2009). There are numerous advantages for companies that incorporating social networking can provide that relate to coordinating marketing, customer services and tracking feedback worldwide on their products and services (Saluja & Singh, 2014). Some of the MIS efficiencies using SNs that AADC can benefit from include:

- EDI (Electronic Data Interchange)
- Transmissions over VANs (Value-Added Networks)
- Company interfaces; various IT software application packages
- Internet Service Providers (ISPs)-for competing in the global market
- Communication in firms and with suppliers to improve supply chain efficiency
- More advanced IT software, hardware; global network access; innovative website design that would appeal to international consumers worldwide to increase customer base
- Developing IT value chain system flows is a major part of how successful e-commerce will be for business Value-added supply chain of distribution links information, communication, services and products between the business and customers

- Businesses bond together through IT streamlining of services
- Online transaction processes; for improved customer service and online shopping
- IT-enabled MIS within the firm that would link to all vendors and suppliers along the global supply chain
- Competitive MIS information on all companies linked within their specific industry, including statistical data and offline and e-marketing strategies
- Increased global management, employee communication, productivity, savings
- Integrated Marketing Communication (IMC) marketing strategies
- Focus on advertising, public relations, direct marketing and consumer and trade promotions to increase consumer awareness and company sales revenues and to better satisfy international customer needs (O'Brien, 1999).

METHODS AND DATA

Using SM was never part of AADC yet. So it might be difficult to conduct a qualitative methodology which is all about running interviews with experts and so on. As a result of the factors discussed previously, the quantitative methodology is seen to be the best in this situation. Furthermore, data, in this research, is going to be collected from two sources; secondary and the primary one.

This project proposes a research design that includes mixed methods of both primary and secondary data collection methods to provide a comprehensive approach to collecting and analyzing data. In order to collect data required, a questionnaire survey will be applied among AADC IT department containing fifteen questions distributed by email to current AADC employees. Questionnaire will involve various questions topics with social networking, MIS, training and other related topics. There are also recommendations from these employees to provide important data on MIS social networking strategies AADC can focus on in the future and what they are planning to implement for upgrading their online customer services and overall organizational performance.

In the other hand, secondary data methods will be involved in the research from an academic literature review of many of the latest online journals, textbooks and essays dealing with the subjects of social networking and MIS. It also describes some of the most recent social networking developments worldwide and how they impact organizations. It will detail some of the most popular SNs that many people in UAE use and their benefits for organizations.

These research methods will be appraised using calculated survey results that are analyzed with quantitative research methods for statistical percentages to develop overall conclusions and recommendations for AADC on the subject of MIS social networking strategies helping to upgrade employee and organizational performance. As illustrated in table 1, an arrangement will be made in favor to let AADC employee participate in the survey effectively.

RESULTS AND DISCUSSION

Data resulted from the questionnaire is going to be analyzed with respect to both research questions and objectives. Questions shared in the survey were classified into four main sections: (1) demographic details, (2) Usability of SM, (3) Values and benefits, (4) challenges. More importantly, the population of this questionnaire is 1000 (AADC staff), however, the total number of AADC staff who were asked to participate in the questionnaire is 227.

Demographic Details

Participants were asked about their level of education, age, gender, and department they work at. The value of this part is to correlate age and education and working environment with the rest of questions such as usability observed in the discussion.

Table 1. Summary of information associated with data collection and research instruments

Methods of Data Collection	Research Instruments	Analysis Details	Details of the Analysis
Primary Source	Questionnaire	Evaluation of the impact of SM on the AADC in order to evaluate the benefits and constraints of the SM.	Responses from the AADC users would be used to obtain the required views as per the questionnaire developed for this purpose.
Secondary Source	Books, Demographic Surveys, Government Publications, Journals, Magazines, News Papers, Reports, Scholar Journals, Previous Research Studies, etc.	Statistical analysis of the data and evaluation of cross tabulation of the variables to check the reliability and validity of our research study.	The reports pertaining to the SM and its impact on the development of the AADC productivity and efficiency

Figure 1 shows that more than 70% of the participants were females and almost 30% were males. This observation correlates well with the existing gender ratio at AADC since 75% of staff are females.

Referring to Figure 2, the majority of participants are holding a bachelor degree (51.58%) from both local and foreign universities.

This study also considered participants age as depicted in Figure 3 where more than 50% of the participants were aged between 25 and 34.

Figure 4 shows a 27% of the participants are working in customer service division as it is the largest division of the AADC.

Figure 1. AADC Employees gender

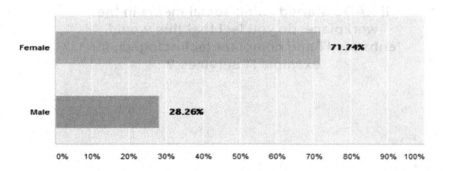

Figure 2. AADC Employees level of education

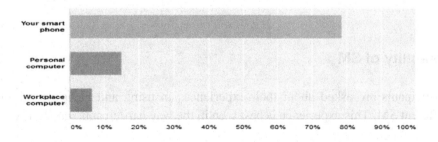

Figure 3. AADC Employees age

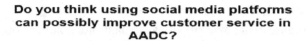

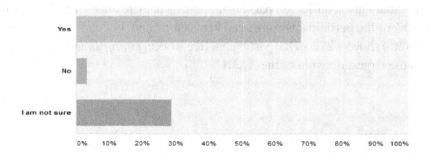

Figure 4. Departments participating

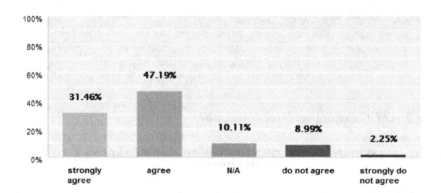

Usability of SM

Participants are asked about their experiences in using and navigating through different SM. This experience is best seen in the way participants' preferences, the main reason these SM are accessed, length of the period they may spend to get what is needed and eventually overall experience.

Figure 5 depicts that the majority of AADC staff tend to use YouTube. Twitter came in the second position.

In this study, there was attempt to highlight the reasons why AADC staff use SNs. Figure 6 shows that most of the staff are using SN to search about the organization's policies such as promotions, policies, how to use covenant ways of payments and recruitments. Only a small category of the respondents use SN to make new friends who never met, discuss topics with experts and posts their topics for comments.

Figure 5. Preferred social networks among AADC employees

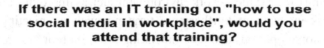

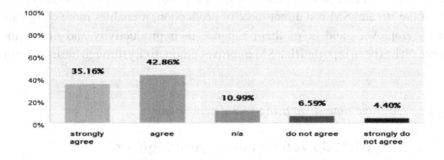

Figure 6. Reasons why AADC employees using social networking

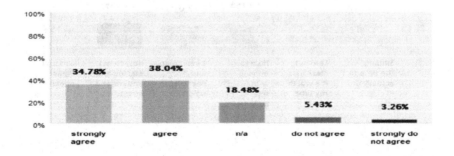

Figure 7 shows that most of the AADC staff see that most of the information they are looking for can be easily found within period of (0 – 30) minutes. This information is not only easy to find, but it is they are richer in content and varied in sources.

Figure 8 depicts most of AADC staff feel positively about using SM on daily basis and they welcome the idea of incorporating this technique in AADC framework. However, AADC staff would feel more comfortable if issues like piracy and photography are eliminated in the adopted SN.

Values and Benefits of SM in AADC

In this part of analysis, there is a need to elaborate on benefits noted by employees and how SM is going to add value to work environment such as improving efficiency, and employee work engagement. Furthermore, time management, meeting experts, discussing professional opinions and recommending the right training are some of many values recognized and could be helping the productivity of the company at the end of the stream. SM is a direct force of production, it enables more competition among employees and helps them enhance their productivity and makes them happier. It has been reported that SM improves productivity through the use of online

Figure 7. Average time spent using social networking

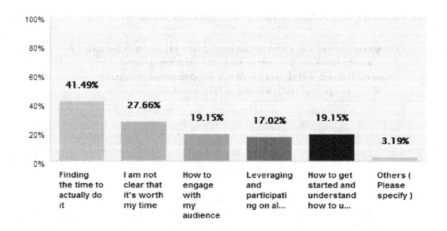

Figure 8. Overall experience with using SM

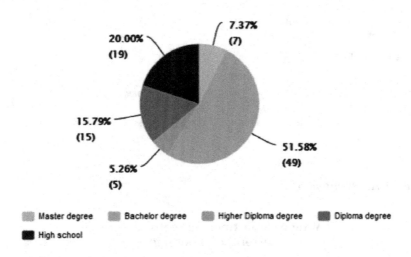

What is the highest level of education you have completed?

*For a more accurate representation see the electronic version.

collaboration rooms. Figure 9 illustrates that over 60% of staff who feel confident to some extent about an organization using SM and feel it would enhance their experience if they had awareness sessions showing values and benefits in global manner and how this technology is going to serve and fulfill company needs and help employees accomplish their work efficiently and effectively.

Figure 10 indicates that smart phones are the most popular among AADC employees. Applications provided from Android and ISO and other smart phone operating system allows you to follow users and search for information.

Apparently as illustrated in Figure 11, most of the participants feels optimistic about SM and there will be a huge retention on customer service as proven in many examples. Moreover, participants describe that this retention could improve the organization's reputation.

AADC employees believe strongly that by implementing SM with IT services framework and other supporting technologies will be in a better shape. Figure 12 shows 47.19% agree with discussion of implementing social networking websites. However, other believe that polices should be revised and adjusted to reduce the overall system's vulnerability and avoid being hacked.

Figure 9. Overall experience with using SM

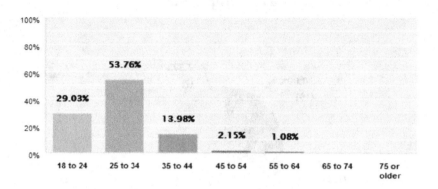

Figure 10. Accessing to SM

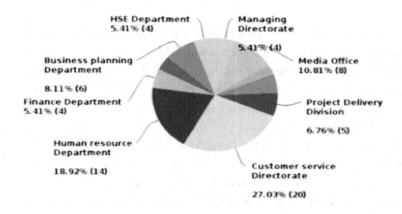

Challenges and Barriers to SM in AADC

In this part of data analysis, participants' feedback is analyzed with regard to different challenges. As many AADC employees agree with idea of implementing SM, there was a need of training them to enlighten them on how to best use and

Figure 11. Improving customer service in AADC

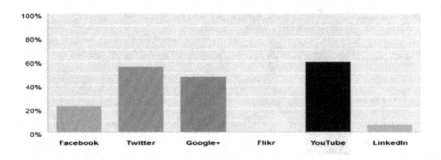

Which of the following social media website do you prefer (You may choose more than one answer)

Figure 12. Enhancing technologies and IT services

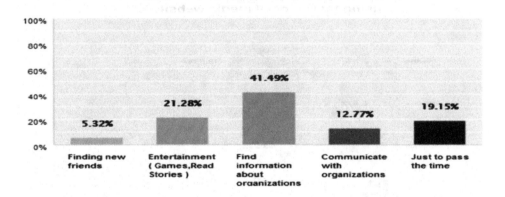

What is the main reason you use these social networking websites for ?

take full benefits from SM. Figure 13 reveals that almost 80% of the employees are willing and excited to have training on how to use SM in a working environment.

Utilizing social networking in operations is accepted and welcomed by most of the participants. Moreover, having new thoughts applied by different organizations and published in Facebook or Twitter will open a new era to enhance rules, polices, and processes. Figure 14 shows that staff who have experiences in social networking websites agreed that there is a strong relationship between social networking websites and enhancing processes.

Figure 13. Providing SM training

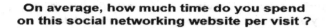

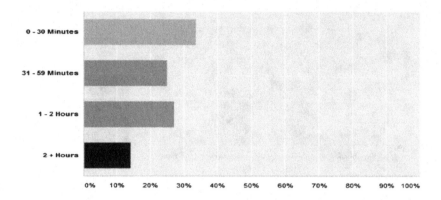

Figure 14. Utilizing social networking in operations

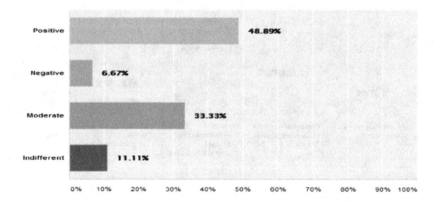

Number of staff specified that the hugest challenge to SM is not having time to actually do it, but this is because either management is not convinced yet about the importance of SM and its positive impact on the organization, so it does not encourage its staff to learn about it, or probably the employee does not manage his/her time in a proper way to use the SM effectively. In addition and by referring back to Figure 7 which illustrates that most employees spent usually between (0 – 30) minutes on SNs/SM and hence time management for the employees is an element to explore.

Figure 15. Challenges to SM

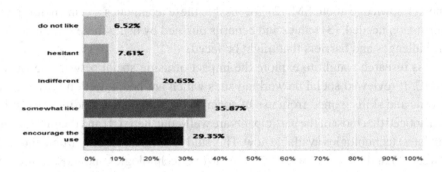

LIMITATIONS AND FUTURE DIRECTIONS

The research limitations were that the organization is owned by the government so there is certain data which was not accessible to obtain the full information needed to extensively study the topic. Although the internet provides detailed information on social networking and lists many different types available for organizations to use, the impact that these online platforms have had on the company could have been more suitably described if the employees were not worried that their survey answers may get them into trouble with their supervisor. Some directions for further research would be to evaluate how social networking has the potential for providing future global partnerships for the government and its subsidiaries.

Social networking could be examined for its overall benefits in the future once the government is better equipped with HRM training programs. This would allow for conducting IT training for Emiratis in English so that they could track online global SN activity related to learning about performance improvement. Future research opportunities proposed in this work are as follows:

- How can organizations use social networking to develop their weaknesses and convert them into strengths?
- In what ways can social networking become a part of the daily business routine for organizations?
- How can companies better utilize social networking so that it helps them track online customer services?

17

CONCLUSION

SM has many impacts on the business. To measure these impacts and how these impacts influence on productivity, this work had to set 4 different objectives: (1) general knowledge about SM, (2) how usable these techniques are and helpful with information needed, (3) values and benefits pursued by using these techniques and (4) challenges and barriers that might be faced.

This research sought to explore the impact of using social networking sites in AADC. It reviewed social networking sites which are being used in business, the benefits and skills gained from use of social networking sites. Additionally, it has been noticed that most of the participants are well-educated staff and they are familiar with latest technologies available now. This study shows that most participants aged between 25 and 34 are familiar with the current technologies and most of them fit also into their organization in general and to their work environment in particular. The results of survey also show that customer service division has the knowledge about these technologies and its associated impact since staff in this division are always required to provide customers satisfactions surveys which revealed that customers wish to have a channel over social websites.

The results of the survey came very supporting to the literature review, for example YouTube is the most famous SM among employees, which corresponds to what was stated by Ingalls (2012). This study reveals that "YouTube" is one of the popular social sites for downloading music and video. It can be clearly seen through survey results that the main reason staff access SM is to find more about organization information and feeds. Since Facebook and Twitter have become so popular, most of governmental organizations have their own pages on these SM as they regularly broadcast their news and offer/advertise their services there. Most of the staff also spent only between (0 – 30) minutes checking the feeds over their social accounts to find out who liked their photos and/or entered comments.

Challenges, defiantly, are one of the research main topics which give insight into points needs to be focused on and studied well, so whenever those challenges come as real, the plan to manage those challenges and barriers is ready to be implemented. One of those challenges are providing well-planned training for staff, which covers all core competencies in order to have the maximum benefits of an IT solution. In this research, many of participants wish to have the proper training, which enable them to deal with the majority of SM aspects such as searching for information about specific events such as marketing campaigns over a Facebook or Twitter.

Additionally, one of the challenges which observed during the survey is to what extent AADC staff believes that AADC can fully utilize SM to modify their business processes. The social networking is a new revolution, so it is quite understandable to have some doubts and worries from staff about how good/bad that can be. By extending the boarder of SN awareness among AADC staff and engaging them in all phases (gathering requirements to implementations), knowledge among staff will be shared and a common understanding is ensured.

Lastly, because AADC staff are busy with their daily tasks and management does not recognize SM as it should be, staff may face quite hard time to find the proper timing to practice and navigate into SM websites. Some of them do not have access to the internet, because their line managers want them to focus just to finish their duties and daily tasks.

This study helps to provide advice and recommendations based on the study outcomes and data analysis. Here are some recommendations for AADC to upgrade their social networking capabilities to increase communication in the workplace, customer and employee satisfaction levels, and overall efficiency in establishing an online presence:

1. Accessibility to social networking sites in the workplace.
2. Develop specific interest social networking objectives for employees.
3. Allow SNs in different languages.
4. Allow various activities to be conducted on social networking sites.
5. Create online forums as an efficient platform of putting employees and customers together to share information and gain feedback.
6. Develop links from the company website to different types of SN sites to gain added value and benefit from the numerous business network opportunities available online.
7. Launch awareness courses about how useful SM can be and what values organization would acquire.
8. Nominate SN trainer for the customer service division.
9. Create awareness and training courses over YouTube and make them available to staff.
10. Give employees assignments to let them check for events over social networks. Then ask them to discuss their findings together and eventually propose a convenient way to implement such event in a similar way.

Managing customer service through SM has the ability to convert complaints into future revenue, therefore it is imperative to investigate and analyze all types of customers' feedback especially through SM. Taking into consideration the current status of SM and how its users are becoming very demanding where the speed of response is becoming critical. Customers expect a prompt response and hence organizations should not treat SM tickets like any other standard ticket. There are various recommendations on how to set a priority criteria including the following:

- Critics or negative comments from unhappy customers
- Service or product requests that are urgent
- Issues that affect many customers
- Acknowledging customers who provided feedback

This study did not explore the impact of SM on customers based on their demographic profiles. This is a window of opportunity to extend this work to include this aspect, analyze its results and draw conclusions illustrating the impact of SM on customers based on demographic profiles. In addition, probable areas of expansion for upgradation of customer service through SM including but not limited to correcting poor service, communicate updates on improvements, turn a negative experience into a positive one and be proactive rather waiting for complaints.

REFERENCES

AADC. (2013). *About us*. Retrieved November 14, 2014, from http://www.AADC.ae/en/about-us.aspx

Al-Jenaibi, B. (2011). The Use of Social Media in the United Arab Emirates – An Initial Study. *European Journal of Soil Science, 23*(1), 84–100.

Alreyaee, S.I., & Ahmed, A. (2015). Trends in Social Media Usage: An Investigation of its Growth in the Arab World. *International Journal of Virtual Communities and Social Networking Archive, 7*(4), 45-56.

Angeletou, S. (2006). Information Systems for e-Government - A Case in the Hellenic Ministry of Education. *International Journal of Public Information Systems, 2*(2), 55–68.

Asur, S., & Huberman, B. A. (2010). *Predicting the Future With Social Media*. Retrieved March 22, 2014, from http://www.hpl.hp.com/research/scl/papers/socialmedia/socialmedia.pdf

Bell, J. (2010). *Doing your research project – A guide for first-time researchers*. Buckingham, UK: OUP.

Biagi, S. (2004). *Media Impact*. Wadsworth Publications.

Boyd, D. M., & Ellison, N. B. (2007). Social Network Sites: Definition, History, and Scholarship. *Journal of Computer-Mediated Communication, 13*(1), 210–230. doi:10.1111/j.1083-6101.2007.00393.x

Castells, M. (2009). *Rise of the Network Society*. Blackwell Publishers. doi:10.1002/9781444319514

Comer, D. (2009). *The Internet Book*. Prentice Hall.

Creswell, J. (2012). *Educational research: Planning, conducting, and evaluating quantitative and qualitative research*. Prentice Hall.

Dahi, M., & Ezziane, Z. (2015). Measuring e-government adoption in Abu Dhabi with Technology Acceptance Model (TAM). *International Journal of Electronic Governance, 7*(3), 206–231. doi:10.1504/IJEG.2015.071564

Dyer, P. (2013). The top benefits of social media marketing. *Social Media Today*. Retrieved February 6, 2016, from http://socialmediatoday.com/pamdyer/1568271/top-benefits-social-media-marketing-infographic

Eldon, E. (2009). *2008 growth puts Facebook in better position to make money*. Retrieved February 6, 2016, from http://venturebeat.com/2008/12/18/2008-growth-puts-facebook-in-better-position-to-make-money/

Ezziane, Z., & Al Kaabi, A.A. (2015). Impact of Social Media towards improving Productivity at AADC. *International Journal of Virtual Communities and Social Networking Archive, 7*(4), 1-22.

Ezziane, Z., & Al Shamisi, A. (2013). Improvement of the organizational performance through compliance with best practices in Abu-Dhabi. *International Journal of IT/Business Alignment and Governance, 4*(2), 19–36. doi:10.4018/ijitbag.2013070102

Gentle, A. (2009). *Conversation and Community: The Social Web for documentation*. Fort Collins: XML Press.

Grant, C. (2007). *Uncertainty and communication*. Basingstoke, UK: Palgrave MacMillan. doi:10.1057/9780230222939

Hanafizadeh, P., Ravasanm, A. Z., Nabavi, A., & Mehrabioun, M. (2012). A Literature Review on the Business Impacts of Social Network Sites. *International Journal of Virtual Communities and Social Networking*, 4(1), 46–60. doi:10.4018/jvcsn.2012010104

Haythornthwaite, C. (2005). Social networks and Internet connectivity effects. *Information Communication and Society*, 8(2), 125–147. doi:10.1080/13691180500146185

Herman, E. (2009). The Propaganda Model. *Journalism Studies*, 1(1).

Khanna, S., & Wahi, A. K. (2014). Website Attractiveness in E-Commerce Sites: Key Factors Influencing the Consumer Purchase Decision. *International Journal of Virtual Communities and Social Networking*, 6(2), 49–59. doi:10.4018/ijvcsn.2014040104

McKeen, J. (2011). *Making IT Happen: Critical Issues in IT Management*. Wiley Series in Information Systems.

McNurlin, B. (2009). *Information Systems Management in Practice*. Prentice Hall.

McQuail, D. (2000). *McQuail's Mass Communication Theory*. London: Sage.

Mellenbergh, G. (2008). *Advising on research methods: A Consultant's Companion*. Johannes van Kessel Publishing.

Miranda, F.J., Rubio, S., & Chamorro-Mera A. (2014). Facebook as a Marketing Tool: An Analysis of the 100 Top-Ranked Global Brands. *International Journal of Virtual Communities and Social Networking Archive*, 6(4), 14-28.

Mukherjee, S. (2016). Leadership network and team performance in interactive contests. *Social Networks*, 47, 85–92. doi:10.1016/j.socnet.2016.05.003

O'Brien, J. (1999). *Management Information Systems: Managing Information Technology in the Internet-worked Enterprise*. Boston: Irwin McGraw-Hill.

Robson, C. (2007). *Real-world research: A resource for social scientists and practitioner researchers*. Malden, MA: Blackwell Publishing.

Salem, F. (2009). *Cross-Agency Collaboration in the UAE Government: The Role of Trust and the Impact of Technology*. Retrieved October 22, 2014 from http://www.mbrsg.ae/HOME/PUBLICATIONS/Research-Report-Research-Paper-White-Paper/Cross-Agency-Collaboration-in-the-UAE-Government-T.aspx

Saluja, D., & Singh, S. (2014). Impact of Social Media Marketing Strategies on Consumers Behaviour in Delhi. *International Journal of Virtual Communities and Social Networking*, 6(2), 1–23. doi:10.4018/ijvcsn.2014040101

Stark, H. T., & Krosnick, A. J. (2017). GENSI: A new graphical tool to collect ego-centered network data react. *Social Networks, 48,* 36–45. doi:10.1016/j.socnet.2016.07.007

Van Tubergen, F., Al-Modaf, O. A., Almosaed, N. F., & Al-Ghamdi, M. S. (2016). Personal networks in Saudi Arabia: The role of ascribed and achieved characteristics. *Social Networks, 45,* 45–54. doi:10.1016/j.socnet.2015.10.007

Von Hippel, E. (2005). *Innovation.* Cambridge, MA: MIT Press.

Wikle, T. A., & Comer, J. C. (2012). Facebook's Rise to the Top: Exploring Social Networking Registrations by Country. *International Journal of Virtual Communities and Social Networking, 4*(2), 46–60. doi:10.4018/jvcsn.2012040104

APPENDIX

Survey Questionnaire

Part 1: Demographic Details and the Current IT Governance Framework Used

Q1. What is your gender?
    [ ] Male
    [ ] Female

Q2. What is the highest level of education you have completed?
    [ ] Master degree
    [ ] Bachelor degree
    [ ] Higher Diploma degree
    [ ] Diploma degree.
    [ ] High school
    [ ] did not attend school

Q3. What is your age?
    [ ] 18 to 24
    [ ] 25 to 34
    [ ] 35 to 44
    [ ] 45 to 54
    [ ] 55 to 64

Q4. Which department/directorate are you currently working in?
    [ ] Managing Directorate
    [ ] Media Office
    [ ] Legal Office
    [ ] Electricity Directorate
    [ ] Water Directorate
    [ ] Asset Management Directorate
    [ ] Project Delivery Division
    [ ] Customer service Directorate
    [ ] Legal Office
    [ ] Human resource Department
    [ ] Finance Department
    [ ] Supply Department
    [ ] Business planning Department
    [ ] HSE Department

Part 2: Understanding and Identifying the Usability of Social Networking With AADC

Q5. Which of the following social media website do you prefer (you may choose more than one answer)?

[ ] *Facebook*
[ ] *Twitter*
[ ] *Google+*
[ ] *Flickr*
[ ] *YouTube*
[ ] *LinkedIn*

Q6. What is the main reason you use these social networking websites for?

[ ] *Finding new friends*
[ ] *Entertainment (Games, Read stories)*
[ ] Find information about organizations
[ ] Communicate with organizations
[ ] Just to pass the time

Q7. On average, how much time do you spend on this social networking website per visit?

[ ] 0 - 30 Minutes
[ ] 31 - 59 Minutes
[ ] 1 - 2 Hours
[ ] 2 + Hours

Part 3: Understanding and Identifying the Value and Benefits of Social Networking With AADC

Q8. What was your overall experience while using these social media website?

[ ] Positive
[ ] Negative
[ ] Moderate
[ ] Indifferent

Q9. How do you feel about organizations that have a presence on social networking websites?

[ ] Do not like
[ ] Hesitant
[ ] Indifferent
[ ] Somewhat like
[ ] Encourage the use

Q10. Which device do mostly you use to access social media?
    [ ] Your smart phone
    [ ] Personal computer
    [ ] Workplace computer

Q11. Do you think using social media platforms can possibly improve customer service in AADC?
    [ ] Yes
    [ ] No
    [ ] I am not sure

Q12. If AADC started using social media in the workplace, do you feel that this would enhance IT and computer technologies, for example IT services?
    [ ] Strongly agree
    [ ] Agree
    [ ] Neither
    [ ] Disagree
    [ ] Strongly Disagree

Part 4: Challenges and Barriers With Social Media in AADC

Q13. If there was an IT training on "how to use social media in workplace", would you attend that training?
    [ ] Strongly agree
    [ ] Agree
    [ ] Neither
    [ ] Disagree
    [ ] Strongly Disagree

Q14. Do you believe companies can utilize social networking websites to make their operations and business processes more productive or efficient?
    [ ] Strongly agree
    [ ] Agree
    [ ] Neither
    [ ] Disagree
    [ ] Strongly Disagree

Q15. What do you feel to be your biggest challenge with social media?
    [ ] Finding the time to actually do it
    [ ] I am not clear that it's worth my time
    [ ] How to engage with my audience and interact with them
    [ ] Leveraging and participating on all social media networks
    [ ] How to get started and understand how to use it
    [ ] Others (Please specify below)

Chapter 2

Functional and Non-Functional Requirements Modeling for the Design of New Online Social Networks

Sarah Bouraga
University of Namur, Belgium

Ivan Jureta
Fonds de la Recherche Scientifique, Belgium & University of Namur, Belgium

Stéphane Faulkner
University of Namur, Belgium

ABSTRACT

Online social networks (OSNs) such as Facebook and LinkedIn are now widely used. They count users in the hundreds of millions. This chapter surveys popular social networks in order to present a pattern of recurring functional requirements as well as non-functional requirements, and a model of that pattern in the i* requirements modelling language. The pattern can serve as a starting point for requirements engineering of new OSNs. The authors test their model by applying it to a popular OSN, namely Twitter.

DOI: 10.4018/978-1-5225-5715-9.ch002

INTRODUCTION

Online Social Networks (OSNs) have grown to be popular over the last decade. They are systems that allow users to share content about any subject they wish, with anyone. To become a member, one has to create an account, and establish relationships with other users. OSNs allow then their users to connect online with their friends, communicate with each other, and share content. The most popular OSNs, such as Facebook, Twitter, or LinkedIn count hundreds of millions of members.

Considerable research has been carried out about OSNs. The literature mostly focuses on the structure of the networks, their topological characteristics, or on the user activity. Also, several studies explore the privacy and trust issues. Indeed, the OSNs have sparked a lot of interest, in part because of the volume of information they gather about their users. The latter give away a considerable amount of information (consciously and unconsciously) through their activity: their profile, their posts, their friends, their browsing activity, etc. This has led to various concerns.

However, as far as we know, researchers have not explored the patterns of features of the OSNs. What kind of features does an OSN have? What are the recurring features that we can find in several OSNs? Does a pattern of features exist that would be shared by OSNs? How important the features of OSNs are to users? Not much attention has been invested in Requirements Engineering (RE) for OSNs, either. What should be taken into account when designing an OSN? How should we model the recurring features? Can we define a requirements pattern for the design of OSNs?

We believe that these research questions are valuable because RE is one of the most important phases in the design of any Information System (IS). It is the first stage in the development of an IS, when the purpose of the IS is decided. RE is concerned with the elicitation, evaluation, specification, consolidation, and evolution of the objectives, functionalities, qualities, and constraints a software- based system should meet within some organizational or physical setting (van Lamsweerde, 2008).

In this paper, we propose to model a pattern of OSN features, using an existing RE modeling language, namely i* (Yu, 1997). More specifically, the goals and the corresponding contributions of this paper are twofold. Firstly, we seek to identify a pattern of features of OSNs. Secondly, we will model these features using an existing requirements engineering language, namely i*, and evaluate if it can represent accurately all the features present in the pattern.

In order to address these questions, we apply the following research methodology. Firstly, we identify features of various OSNs and compare them across those systems; as opposed to features specific to one particular OSN. Secondly, we analyze this list of features with regard to the selected OSNs: are the given features present in the OSN? Which form does the feature take? We will see if a pattern emerges from the discussion. We also discuss the implications of our findings for the design of future

OSNs; and more specifically, the implications for the RE phase in the design of an OSN. Thirdly, we apply our modeling pattern to a specific OSN, namely Twitter. If a pattern of features common to several OSNs is found, we will know that one should pay attention to these particular features when designing an OSN.

The rest of this paper is structured as follows. First, Related Work is discussed in Section 2, as well as a brief summary of several popular OSNs. Then, we introduce the features we identified of OSNs, and analyze them in relation to the selected OSNs, in Section 3 Section 4 respectively. In Section 5, we turn to the modeling of the features using an existing RE language, namely i*; and we discuss the results in Section 6. Finally, Section 7 concludes the paper.

RELATED WORK

General Literature Review

Let us first consider a definition of OSN. Boyd et al. (2007) define social network sites as:

Web-based services that allow individuals to (1) construct a public or semi-public profile within a bounded system, (2) articulate a list of other users with whom they share a connection, and (3) view and traverse their list of connections and those made by others within the system. The nature and nomenclature of these connections may vary from site to site.

Another definition was provided by Kaplan et al.; they state that "Social Media is a group of Internet-based applications that build on the ideological and technological foundations of Web 2.0, and that allow the creation and exchange of User Generated Content" (Kaplan & Haenlein, 2010).

The structure of OSNs have been studied by various authors. The structure of specific social networks was explored: the structure of Flickr, YouTube, LiveJournal and Orkut (Mislove et al., 2007); the structures of Cyworld, MySpace, and Orkut were compared, in order to analyze the topological characteristics of these sites (Ahn et al., 2007); the structure of two large Chinese OSNs, namely Sina blogs and Xiaonei SNS, were studied (Fu et al., 2008); the structure and evolution of Flickr and Yahoo! 360 (Kumar et al., 2010); and the topological characteristics of Twitter were analyzed (Kwak et al., 2010). Also, Vizster was proposed to represent OSNs. The system is used by members of online communities to explore their social network in the way they desire (Heer & Boyd, 2005). Mislove et al. (2008) claim that various OSNs, despite their different purposes, share a number of similar structural features, namely: highly skewed degree distribution, a small diameter, and significant local clustering. Viswanath et al. (2009) explored the user interaction in Facebook. They discovered that the activity distribution is skewed. Leskovec et al. (2010) proposed a method to predict "the signs of links in large social networks where interactions can be both positive and negative". The relationship strength between members of OSNs were explored; and therefore, a model was proposed for the representation and the inference of the relationship strength (Xiang et al., 2010).

Several authors researched the reasons explaining the decision to use an OSN by different categories of people (teenagers Livingstone, 2008), college students (Park et al., 2009) or emerging adults (Subrahmanyam et al., 2008)), as well as the user activities in OSNs. The reasons why people are members of and use an OSN are: users can satisfy a "friend" need, as well as the need for a source of information (Raacke & Bonds-Raacke, 2008); emerging adults use OSNs "to connect with others, in particular those in their offline lives" (Subrahmanyam et al., 2008); a study researching the reasons why people microblog was carried out (Java et al., 2007). The user activity was also explored by several authors. Guo et al. (2009) studied the user activity in the knowledge-sharing oriented OSNs (as opposed to network-oriented OSNs), specifically: the participation of members in the network, the generation and posting of content, and the quality of content. Schneider et al. (2009) proposed a methodology for the identification of OSN sessions, as well as the user activities within an OSN session. Benevenuto et al. (2009) observed that the user activities in Orkut can be categorized in one of the following features: Universal search, Scrapbook, Messages, Testimonials, Videos and Photos, Profile and Friends, Communities, and Other.

The user profile has also been studied in the literature: the role of profile elements in the creation of online friendship connections (Lampe et al., 2007); how "a social network profile's lists of interests can function as an expressive arena for taste performance" (Liu, 2007); the kinds of information Facebook users shared on their profile (the study provides a comprehensive checklist identifying these types of information) (Nosko et al., 2010). Mislove et al. (2010) also explored the possibility to infer the attributes of some users in an OSN, given the attributes of some users in the same network. The authors discovered that members of social networks usually share the same attributes as their friends; and thus the attributes of some users can be inferred accurately given the attributes of other users belonging to the same community. The authors also proposed their own approach to detect communities for multiple attributes.

The privacy issue was examined by several studies: the trust and privacy issues were compared between Facebook and MySpace (Dwyer et al., 2007); the "concerns and strategies of users [...] on Facebook" (Strater & Lipford, 2008); "NOYB, an approach that provides privacy while preserving some of the functionality provided by online services" was proposed (Guha et al., 2008); the relationship between trust and profile similarity was studied (Golbeck, 2009); and the privacy attitudes intentions were measured and compared "against the privacy settings on Facebook" (Madejski et al., 2011).

Several behavior-related studies were carried out: "The Spread of Behavior in an Online Social Network Experiment" (Centola, 2010); a human behavior-oriented analysis of Club Nexus, an online community at Stanford University (Adamic et al., 2003); and the relationship between college students' use of social network (here Facebook), and their stock of social capital was explored (Valenzuela et al., 2009).

Specific OSNs were also studied: MySpace (Caverlee & Webb, 2008); Facebook (Lampe et al., 2006; Pempek et al., 2009); YouTube (Lange, 2007); and Massively Multiplayer Online Games (MMOGs) (Ducheneaut et al., 2006).

The cited research differs from our work here, despite some similarities. We seek to identify and model a pattern of recurrent features of OSNs, so we are not concerned with the formal properties of graphs induced by user relationships. We are interested in the reasons why people use OSNs, the activities they engage in, the user profile, and the privacy issue; but not in the sole goal of better understanding these matters. Instead we are interested in these OSNs related concepts because they are useful in the RE phase. Indeed, the latter should thoroughly analyze the situation before and after the introduction of the system-to-be (here an OSN); hence the RE step should consider the reasons that motivate people to become an OSN member, it should consider their activities, the way they will present themselves, and the issues the users could meet while using the system.

Software Engineering for Online Social Networks

Literature on the software engineering, including RE, aspect of OSNs is rare. Some researchers focused their work on the privacy requirements: Sheth et al. (2014) and Van Der Spye & Maalej (2014) proposed some guidelines for developers when dealing with privacy requirements; and Gurses et al. (2008) proposed privacy heuristics for OSNs.

Various authors focused on the design of a specific OSN. Examples include Gospill et al. (2013) Lai et al. (2010), Plotnik et al. (2009), Chao et al. (2011), and Magoutis et al. (2015). Gospill et al. (2013) proposed a "Social Media framework to support Engineering Design Communication". They elicited the requirements for the application of Social Media through the review of the literature and by analyzing the suitability of Social Media for the support of Engineering Design Communication (EDC). A framework for the development of a social media tool for the support of EDC was proposed. The framework is composed of three elements:

(i) a Communication Process; (ii) an EDC classification matrix; (iii) a set of tables. Lai et al. (2010) addressed the design and implementation of an OSN with face recognition. The authors proposed the following system architecture, but did not explicitly identified the requirements for the development of the system, nor explained how to elicit them. The system architecture consists of three elements: (i) User database, (ii) Face recognition Web service; (iii) Friendship algorithm. Plotnik et al. (2009) proposed the design of an OSN for emergency management. They identified the requirements through surveys. Therefrom, they derived design principles for developers. Magoutis et al. (2015) introduced the design and implementation of a social networking platform for the development and operation of IS. Chao et al. (2011) proposed, and described the development and features of, an interactive social media-based learning environment. Zhang et al. (2014) reviewed the construction, analysis, and applications of Developer Social Networks. The latter are used to facilitate the software engineering tasks.

Other authors focused their work on the issue of the design, implementation and evaluation of more general OSNs.

Kietzmann et al. (2011) proposed a framework defining social media by using seven building blocks, namely: identity, conversations, sharing, presence, relationships, reputation, and groups. They also proposed various recommendations for firms that want to develop strategies regarding social media activities. They gathered these recommendations under the designation "The 4-Cs", indicating that a firm should: Cognize, (Strive for) Congruity, Curate, and Chase. Pereira et al. (2010) revised this framework, and discussed how it can be used in the design and study of social software.

Similarly, Weiss et al. (2013) proposed six recommendations for the "design, implementation and evaluation of social support in online communities, networks, and groups." These recommendations are:

1. Design:
    a. "Address the interdependence between online support and real-world support"
    b. "Address the individual's existing social networks (e.g. family, friends, and co-workers)"
    c. "Target community-wide outcomes and participation of local community groups"
2. Evaluation:
    a. "Adapt and/or develop evaluation measures of support specific to online environments"
    b. "Consider all units of analysis (from interpersonal to community-wide measures of support)"

3. Design, Implementation, and Evaluation:
   a. "Employ ecological systems theory and principles of community-based participatory research"

Karampelas (2013) discussed the technological and methodological aspects of the design of social network platforms and applications. More specifically, the author addressed various phases in the development of a social network platform: the design, the implementation, and the evaluation phases.

Ghafoor & Muaz (2016) proposed a methodology for the design of an OSN using the analysis of human interactions.

Our research differs from these studies, because we do not focus on a specific type of requirements, namely privacy requirements, but instead on all potential features an OSN can propose. However the goal is similar in the sense that we are trying to offer some guidelines to developers, by prioritizing OSN features. Also, our work here is more modest than the ones proposed in (Kietzmann et al., 2011; Pereira et al., 2010; Weiss et al., 2013; Karampelas, 2013), because we are focusing only on the design of new OSNs, and we do not address the evaluation nor the implementation. Also, we are trying to improve existing OSNs by incorporating a RS to them, and we do not model only the existing features proposed by OSNs.

Online Social Networks

In this subsection, we will briefly go through several popular OSNs. In the articles reviewed above, we encountered various OSNs: Facebook, Bebo, MySpace, Orkut, LinkedIn, Meetup, Friendster, YouTube, Flickr, Picasa, LiveJournal, Blogspot, Hi5, Studivz, Tribe.net, QQ, Cyworld, Skyrock (formerly Skyblog), Classmates.com, Asianavenue, Migente, Blackplanet, Tumblr, Pinterest, and Twitter. Among these, we will select the following:

- **Facebook (www.facebook.com):** Is a tool allowing users to "connect with friends and the world around you" (Facebook). Members set up a profile with information such as demographics and interests. They can then connect with people (Friends), through a reciprocated relationship. Users can also post photos, videos, messages; they can join groups; and they can send messages to their friends (Golder et al., 2007).
- **Flickr (www.flickr.com):** "Is a photo-sharing site based on a social network" (Mislove et al., 2007). The "sharing, retrieval, navigation, and discovery of user-contributed images" are based on tags. Users can upload their personal photos online and these are made publicly viewable by Flickr. Flickr also offers various communication tools: create networks of friends, join groups,

send messages, comment on photos, tag photos (users can tag both their personal and others' photos), choose their favorite photos, etc (Marlow et al., 2006).

- **LinkedIn (www.linkedin.com):** Is a professional-oriented social network. Users are encouraged "to construct an abbreviated CV and to establish connections". Tools offered by LinkedIn to its members include: the solicitation of recommendations from other members; privacy settings; the explorations of the direct connections of their connections; the formation of groups; the sending of messages (Skeels & Grudin, 2009). LinkedIn uses a "'gated-access approach', meaning that connecting with others requires either a pre-existing relationship or the intervention of a mutual contact" (Papaharissi, 2009).

- **MySpace (www.myspace.com):** "Is an online community that early adopters helped shape into a music- friendly place where hipsters, indie bands, and fans could network and socialize with one another" (Liu, 2007). MySpace users own a profile gathering information about them. Friend relationships connect user profiles together. Various messaging mechanisms are also offered by MySpace. MySpace members are allowed to decide whether they want their profile to be public or private, that is only accessible to Friends (Ellison et al., 2007; Caverlee & Webb, 2008). MySpace offers features such as forums, and user groups.

- **Pinterest (www.pinterest.com):** Is a growing OSN. The site "revolves around the metaphor of a 'pin board'". When a user finds a photo or an item on the web she likes/finds interesting, she can pin it. All the pinned photos can then be organized into topical collections (fashion, beauty, sports, DIY, etc). Members can "follow one another, repin', like, and comment other pins" (Gilbert et al., 2013). The collections (the boards) can be constantly updated, shared, and are publicly viewable (Dudenhoffer, 2012).

- **Tumblr (www.tumblr.com):** Is a microblogging platform. The site allows its users to post photos, videos, music, texts; to tag their posts; to follow other members; to reblog, like or comment other users' posts; to send and receive messages; to browse for posts.

- **Twitter (www.twitter.com):** Allows users to share status, messages, called tweet (up to 140 characters) with their Followers, about any topic. Users can follow other users, and also post photos and videos. Unlike many other OSNs, the connection between users can be unidirectional, that is unreciprocated (Kwak et al., 2010; Java et al., 2007).

- **YouTube (www.youtube.com):** "Is a popular video sharing site that includes a social network" (Mislove et al., 2007). YouTube was founded as a site enabling users to easily share video content. Users can upload videos and tag them. The site uses those tags to provide users with a list of related videos (Gill et al., 2007). Each of these online social networks will be then used to classify the features we will identify in the next Section.

Features

We extract features from the 8 selected OSNs and we organized these features in categories. We know from Boyd et al. (2007) that a social network is characterized by a Profile and Relationships. We propose here to add others characteristics: Content, Privacy, Goal, Recommendation, and Connection to Other OSNs.

- **Profile:** According to Lampe et al. user profiles constitute "an integral part of social network sites" (Lampe et al. 2007). The profile of a user constitutes a way for the user to present himself to the other members of the social network. We can identify various types of information a user can communicate in his profile:
    - **Login Information:** Email, and username
    - **Identity:** Name, birthdate, address, profile picture
    - **Occupation:** School, job, etc.
    - Family
    - **Beliefs:** Political, religious, etc.
    - **Skills:** Languages, qualifications
    - **Hobbies/Favorites:** Sports, culture, favorite movie, favorite quote, etc.
    - **Else:** About You
- **Relationship:** The relationship can be defined as the link between two users. The relationship can be Unidirectional, or Bidirectional; that is, it can be unreciprocated or reciprocated, respectively.
    - **Unidirectional:** The relationship is unreciprocated. A user likes, or subscribes to a fan page; or follows another user.
    - **Bidirectional:** The relationship is reciprocated. A friend request is sent by a user to another user. The latter has to accept or deny the friend request. If he confirms it, then the relationship is created. Otherwise, no link exists between the two users.
- **Content:** Users can post, and share different types of content. Members of an OSN can share: a status, notes, photos, videos, links, comments, or messages. Members can also join groups.
    - **Text:** A user can post strings of characters. It can be a status, a text, a note, a link to another webpage, quotes, etc.

- ○ **Comment:** A user can comment on his own post, or on another user's post. This post can be a photo, a status, a note, or any other kind of activity.
- ○ **Like/Repost:** A user can like or repost his own posts or another user's. This post can be a photo, a status, a note, or any other kind of activity.
- ○ **Tag:** A user can tag his own post or another user's. The tag can concern users on photos, videos, status. Or the tag can be used to document, and be able to identify a media.
- ○ **Media:** A user can post, comment, like a photo or video.
- ○ **Message:** A user can send a message to another user.
- ○ **Groups:** A user can create and join a group.
- **Privacy:** Users can decide what they want to share with the other users, with their friends, and what they want to keep private. We can thus identify three degrees of privacy: Private (information, or any type of content the user wants to keep for himself); Semi-Public (information or content the user wants to share with a certain type of public, usually his Friends); and Public (information or content the user wants to share with anybody: other users and non-users of the network).
  - ○ **Private:** If a content or information is private, then only the given user has access to it
  - ○ **Semi-Public:** The user decides to share content and/or information with particular groups, or particular categories of users. This group can consist of his friends, friends of friends, groups of friends, etc.
  - ○ **Public:** Anyone has access to the posts shared by the user. It can be every member of the OSN, or even anyone visiting the OSN.
- **Recommendation:** Based on the information shared by the users, as well as on the information inferred from it; OSNs can provide recommendations to its members. These recommendations can be about other users (the Friend suggestion offered by Facebook, for instance); about groups they might want to join; artists they would like, etc.
  - ○ **Users:** A user is recommended a list of people he may know.
  - ○ **Business/Celebrities:** A user is recommended a list of businesses or celebrities she may like.
  - ○ **Content:** A user is recommended photos, videos, social network groups he may like.
- **Connection to Other OSNs:** A connection can be established between various social networks. The link between two different sites can exist to sign in, or share content.

- ○ **Sign In:** A user can sign in to a social network, using his account of another social network.
- ○ **Share Content:** A user can post simultaneously on various social networks.

This is summarized in Table 1.

Classification of the Features by Online Social Network: Their Presence and Their Form

We will now turn to the analysis of the features for each of the 8 OSNs mentioned above.

We start with the Profile features. Each OSN demands a valid email address and a password as login information. Some OSNs allow the use of a phone number (Facebook) or username (MySpace) as login. Also, Flickr and YouTube require the use of a specific email address, namely a Yahoo ID and a Google email address respectively. When setting up their profile, users can provide several pieces of information regarding their identity. Their name and a profile picture are a constant for all OSNs considered here. The other elements are: gender, address/location (Facebook, Flickr, MySpace, Pinterest); birthday (Facebook, Flickr, MySpace); relationship status (Facebook, Flickr), phone number (Facebook), title (LinkedIn); account type (personal, musician), ethnicity, body type, height (MySpace); website (Pinterest, Twitter), URL's blog (Pinterest, Tumblr), Google profile (YouTube). Facebook, LinkedIn, and MySpace allow the users to tell more about their occupation: the school(s) they attended, and the jobs they have (had). Facebook enables the identification of one's family members. A user of Facebook or MySpace can state

Table 1. Features and their categories

Categories	Features	Categories	Features
Profile	Demographics Occupation Family Beliefs Skills Hobby's "About me" text	Content	Text Comment Like/Reblog Tag Media Message Group
Relationships	Unidirectional Bidirectional	Connection	Sign in Share
Privacy	Private Semi-public Public	Recommendations	Friend Public figure Content

her religious and political views, or her religion respectively. With Facebook and LinkedIn, a user can share information about her skills: the languages she can speak (Facebook); her qualifications, summary, specialties, areas of expertise (LinkedIn). Facebook, LinkedIn, and MySpace allow their users to tell more about their hobbies, or their favorite things; more specifically: their favorite quotations, and interests (Facebook); their interests (LinkedIn); their kind of music (MySpace). Finally, Facebook, Flickr, LinkedIn, Pinterest, Tumblr, Twitter, and YouTube leave some kind of an About You space where the user can give a brief description of herself. LinkedIn also allows its members to give advice for contacting them. MySpace gives the chance to its users to tell more about their preferred site, hero, personality, and the reason they use MySpace. A YouTube member can share links on his profile, as well as, feature other channels (see Table 2).

The email address is particularly important for the connection with other OSNs. Other information shared by the user in her profile is used by an OSN to make recommendations to this user. For instance, if the user is 15 years old, she will receive recommendations of pages or content appropriate to that age and it will probably be different from what a 50-year-old will receive. Another example would be a suggestion generated by the OSN based on the hobbies of the user. If a user informs that she likes to play tennis, she might get recommendations about Roger Federer's fan page; while a football fan will be recommended the Cristiano Ronaldo's page.

The relationships between users are one of the characteristics of OSNs, according to Boyd et al. (2007). We can identify two types of relationships: unidirectional and bidirectional. Most of OSNs support the latter. Indeed, Facebook, Flickr, LinkedIn,

Table 2. Profile features classification

	Login	Identity	Occupation	Family	Beliefs	Skills	Interests	Else
Facebook	Email, Phone	Demographics, Picture	School(s), Job(s)	X	Religious, Political	Languages	X	About me
Flickr	Yahoo ID	Demographics, Picture						Describe yourself
LinkedIn	Email	Demographics, Picture	Industry, Experience			Summary, Expertise	X	Personal details, Contact me
MySpace	Email, Username	Demographics, Picture	School(s), Company/ies		Religious		X	Reasons you use MySpace
Pinterest	Email	Demographics, Picture						About you
Tumblr	Email	URL, Picture						About you
Twitter	Email	Demographics, Picture						Bio
YouTube	Google email	Google profile						Description, Links

MySpace and YouTube offer the possibility to their users to send friend requests, and accept or deny them. Other OSNs implement unreciprocated relationships: Pinterest, Tumblr, and Twitter allow users to follow boards, blogs, and twitter accounts respectively. The direction of the relationship is thus unique; even though it can be bidirectional if, for instance, two Twitter users follow each other. We should note that Facebook, Flickr, LinkedIn and YouTube also enable their members to like fan pages, follow companies, and subscribe to channels without any reciprocation required (see Table 3).

The type of relationship between users has its importance for several other features: the comments, the likes, the repost, the message, and the privacy. Depending on the privacy set on the post, it could be that only friends are allowed to interact (both send messages and comment/ like/repost posts) with a user.

All OSNs considered here offer the possibility to share some posts and media. Apart from Pinterest, every OSN allows its users to share texts, which take different forms or have different terminologies according to the site. Facebook has the status, notes and links; Flickr has the testimonials; LinkedIn has the recommendation; MySpace has the status and the moods; Tumblr has the texts, quotes, and links; Twitter has the tweet; and YouTube has the discussion. All OSNs allow their users to comment on their own posts or others'. These posts can be a status, a media, or any kind of activity. Flickr and YouTube users can have favorite photos or videos; and MySpace users can share posts. For the other sites, both features (like and repost) are available to the members. We should note that the terminology for the repost can differ from site to site: Facebook, or LinkedIn users can Share; Pinterest users can Repin; Tumblr members can Reblog; and Twitter users can Retweet posts. Users of all OSNs can also benefit from the tag feature. On the one hand, Facebook, LinkedIn, and MySpace allow their users to tag their friends/ connections on photos (Facebook, MySpace), and on status and updates (Facebook, LinkedIn, Twitter); Facebook and

Table 3. Relationship features classification

	Uni-Directional	Bidirectional
Facebook	X	X
Flickr	X	X
LinkedIn	X	X
MySpace		X
Pinterest	X	
Tumblr	X	
Twitter	X	
YouTube	X	X

Twitter also offer the hashtag feature. On the other hand, Flickr, Pinterest, Tumblr, and YouTube users can tag their media to facilitate their identification. As far as the media are concerned, all OSNs offer the possibility to their members to share photos and videos. The only exception is YouTube, where the users can only upload videos. We should also note that Tumblr allows the sharing of gifs. The Message feature is also very popular. Indeed, all OSNs offer this feature. As far as the groups are concerned, Facebook, Flickr, LinkedIn, and MySpace users can create and join groups. Pinterest and Tumblr members can create a board and a blog respectively with multiple contributors. Twitter users can join lists. And YouTube members who subscribe to channels are, in a way, members of a same group.

As mentioned in Table 4, the features related to the content are in close relation to the type of relationships between users, and to the privacy settings. Given the privacy settings, only a group of users could interact regarding the content of the OSN. As far as the tags are concerned, if a user wants to tag another user on a photo, they usually need to be friends.

The various OSNs propose different degrees of privacy to their members. Facebook allows its members to keep content private (only the user can see her personal content), to share it with their friends or a specific group of friends, or to share it with everyone (all Facebook users). Flickr, LinkedIn, and YouTube offer similar privacy settings, but if the content is public then not only all users of the

Table 4. Content features classification

	Texts	Comments	Like/Reblog	Tag	Media	Message	Group
Facebook	Status	X	Like, Share	Friends, On posts	Photos, Videos	X	X
Flickr	Testimonials	X	Favorite	Photos	Photos, Videos	X	X
LinkedIn	Recommendations	X	Like, Share	Friends	Photos, Videos	X	X
MySpace	Status	X	Share	Friends	Photos, Videos	X	X
Pinterest		X	Like, Repin	Photos, Videos	Photos, Videos	X	X
Tumblr	Texts	X	Like, Reblog	Post	Photos, Videos	X	X
Twitter	Tweet	X	Favorite, Retweet	On posts	Photos, Videos	X	X
YouTube	Discussion	X	Like, Favorite	Videos	Videos	X	X

site can see the given content, but non-members too. MySpace offers a different configuration: users cannot keep content or information for themselves. However, they can control with whom they share it: friends only, or users who are over 18 years old, or everyone (including non-MySpace members). Pinterest allows users to keep board for themselves and for whomever they invite. Otherwise, the boards are public, thus accessible to members and non-members of Pinterest. Nevertheless, the number of secret boards a user may have cannot exceed three. Tumblr members only have two possibilities: their posts are either private (only available to them) or public (accessible by everyone, even non-members). Twitter proposes only two options too. The content posted by a user can be either shared with their accepted followers; or it can be shared with the world. Similar to MySpace, a user cannot keep content or information for herself only (see Table 5).

The privacy settings are in close relation with the relationship and content features. Indeed, depending on the privacy set for a specific content; all users, only a portion of users, or a portion of the user's contacts can have access on the given content.

According to Guo et al. (2009), OSNs can be classified in two categories: Knowledge-Sharing and Network-Oriented. We can classify Flickr, Pinterest, Tumblr, and YouTube in the former category; and the others, that is Facebook, LinkedIn, MySpace, and Twitter in the latter. Indeed, for the first group of OSNs, the emphasis is put on the content of the site: photos, music, posts, videos; while for the second group the focus is more on the connections (see Table 6).

All the OSNs considered here provide recommendation to their users (see Table 7). These recommendations are about other users for all of the OSNs. In addition, Facebook, LinkedIn, and Twitter generate recommendations about celebrities or businesses. Furthermore, suggestions about some kind of content are produced by Flickr (photos and videos), LinkedIn (Jobs you may be interested in), MySpace (music, videos, games), Pinterest (boards, pins), Tumblr (posts), and YouTube (videos).

Table 5. Privacy features classification

	Private	Semi-Private	Public
Facebook	X	X	Users
Flickr	X	X	Non-users
LinkedIn	X	X	Non-users
MySpace		X	Non-users
Pinterest	X	X	Non-users
Tumblr	X		Non-users
Twitter		X	Non-users
YouTube	X	X	Non-users

Table 6. Goal features classification

	Knowledge	Network
Facebook		X
Flickr	X	
LinkedIn		X
MySpace	X	X
Pinterest	X	X
Tumblr	X	
Twitter		X
YouTube	X	

Table 7. Recommendation features classification

	Users	Business/Celebrity	Content
Facebook	X	X	
Flickr	X		X
LinkedIn	X	X	X
MySpace	X		X
Pinterest	X		X
Tumblr	X		X
Twitter	X	X	
YouTube	X		X

The recommendations are generated on the basis of the user profile information, and on her activity: her posts, her comments, her browsing activity, her friends, etc.

Finally, all the OSNs offer the possibility to their users to connect, link their account with another OSN account. Facebook offers a connection with several other OSNs, allowing its users to sign in with their Facebook account on MySpace and Pinterest; to find/follow friends on Flickr, MySpace, and Pinterest; and share posts across OSNs, namely Flickr, LinkedIn, Pinterest, Tumblr, Twitter, and YouTube) (see Table 8). Flickr users cannot sign in with another OSN account, but they can share posts simultaneously on Facebook, Google, and Twitter. LinkedIn encourages its members to connect their account with their blogs, or Amazon; and they can also link their account with their Facebook or Twitter to share posts. MySpace users can sign in with their Facebook or Twitter account; they can share status with Twitter; and they can find/follow friends via Facebook. Pinterest users can sign in with

Facebook or Twitter; they can also share posts with Facebook. Tumblr users can share posts on Facebook and Twitter. Twitter users can share posts with Facebook, LinkedIn, MySpace and Tumblr. YouTube members can sign in using Google; and they can share videos using Facebook, MySpace, Twitter, Bebo, Orkut, and Hi5.

This connection feature is linked with, on the one hand the login information since one of the reasons why a connection between two OSN accounts is offered is the signing in. And on the other hand, the connection feature is linked with the content feature, because users can share simultaneously the same post on different sites.

This discussion is summarized in Tables 2 to 8.

MODELING OF THE FEATURES USING I*

Requirements Engineering

Requirements engineering (RE) is concerned with the elicitation, evaluation, specification, consolidation, and evolution of the objectives, functionalities, qualities, and constraints a software-based system should meet within some organizational or physical setting (van Lamsweerde, 20008). The focus of RE is the investigation, delineation and precise definition of the problem world that a machine solution is

Table 8. Connection Features Classification

	Goal		OSN
	Sign In	**Share**	
Facebook	X	X	Twitter, Flickr, LinkedIn, MySpace, Pinterest, Tumblr, YouTube
Flickr		X	Facebook, Twitter, Google
LinkedIn		X	Facebook, Twitter, Amazon, Blogs
MySpace	X	X	Facebook, Twitter,
Pinterest	X	X	Facebook, Twitter, Google, Yahoo
Tumblr		X	Facebook, Twitter,
Twitter	X	X	Facebook, LinkedIn, MySpace, Pinterest, Tumblr
YouTube	X	X	Facebook, Twitter, MySpace, Google, Orkut, Bebo, Hi5

intended to improve. The RE process is an iteration of intertwined activities for eliciting, evaluating, documenting, consolidating, and changing the objectives, functionalities, assumptions, qualities and constraints that the system-to-be should meet based on the opportunities and capabilities provided by new technologies. Those activities involve multiple stakeholders that may have conflicting interests. The relative weight of each activity may depend on the type of project (van Lamsweerde, 20007).

Modeling of the Requirements

As mentioned above, we use i* (Yu, 1997) to model the OSN features. The Strategic Rationale (SR) model allows to express process elements and the rationales behind them. Here, we will not represent the Strategic Dependency (SD) model because we are interested in a higher level of details than the one provided by the SD model. Indeed, we identified generic OSN features, which are implementations that satisfy requirements; and now we want to model these requirements. This requires the level of details allowed by the SR model.

For the purpose of our representation here, we will identify five actors, namely: the user of an OSN, a friend of the user, a celebrity she likes, the public, and the OSN. Figures 2 and 3 represent the requirements and these actors. Because of the size of the model, we used connectors. The model is too large to be represented in one page, we thus broke it into two parts, and the connectors allowed us to connect both parts of the model. The legend can be found in Figure 1.

An internal task of the user is to present herself through her profile. The user depends on the OSN to achieve this goal (modeled by a goal dependency). The OSN will fulfill this goal by displaying to the other users the elements present in the user profile (means-end), that is the features belonging to the profile category: the identity, the occupation, the family members, the beliefs, the skills, the hobbies, else. This is represented by the means-end relationships.

The OSN will depend on the user to provide the information to display, as well as login and password (modeled as resource dependencies).

The user also depends on the OSN to link her with other users (friends and/or celebrities). The task Link users is decomposed into Users be linked uni- directionally and Users be linked bidirectionally. Those are achieved by linking the user to the public figure page, a public, or a private user; and by sending and receiving a response to a friend request respectively. The OSN depends on the user to provide the name of the friend/public figure (resource dependency); and the OSN also depends on the public user and public figure to have an account (resource dependency); and depends on the private user to respond to the follow request to set the uni-directional relationship, or to respond to the friend request, to set the bidirectional relationship (task dependency).

Figure 1. Legend

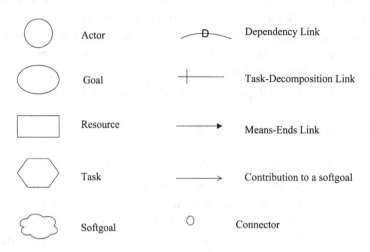

A member of OSN also wants to see and share content. She depends on the OSN to display elements of content. The OSN will achieve the goal "Display content" by displaying to the other users the elements of content shared by the user, that is the features belonging to the content category: text, comment, like/repost, tag, media, message, groups. This is represented by the means-end relationships. The OSN will depend on the user to provide the posts to display (modeled as resource dependencies).

The OSN user manages her privacy settings. She depends on the OSN to achieve this. The OSN offers three possibilities for the management of the privacy, namely private posts, semi-public posts, and public posts (means-end relationships). The OSN also depends on the user to express the level of privacy she wants for each of her post/ piece of information. The OSN also depends on the user to have bidirectional relationships with other users if she wants to have semi-public posts. We can also state that the friends depend on the user to share semi-public posts; while the public depends on the user to share public posts.

The user depends on the OSN to give recommendations. This goal dependency leads to a task Give recommendation, that can be decomposed into: give recommendations about users, give recommendations about businesses/celebrities, and give recommendations about content. We can identify two resource dependencies related to that goal. The OSN depends on the user and her friends for activity (for simplicity, they are not represented in Figures 2 and 3).

Figure 2. Strategic Rationale Model for an Online Social Network - Part 1

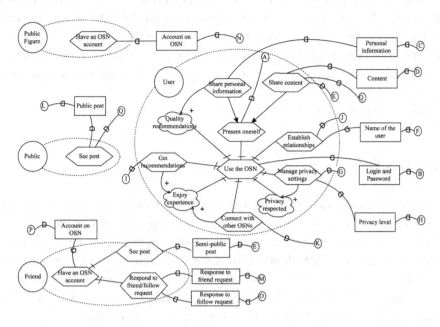

Figure 3. Strategic Rationale Model for an Online Social Network - Part 2

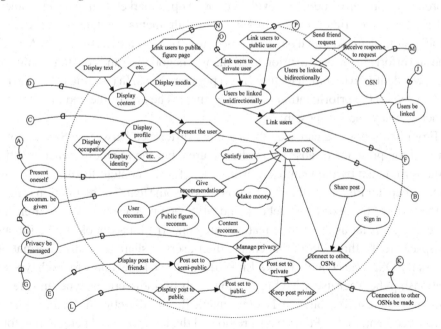

The user also depends on the OSN to connect with other OSNs. This connection with other OSNs is achieved by allowing to sign in with other OSNs; and by sharing simultaneously the same post on various OSNs. The OSN depends on the user to have another OSN account (resource dependency), and to allow the connection across OSNs (task dependency). For simplicity, the latter are not represented in Figures 2 and 3.

This discussion is represented in Figures 2 and 3.

DISCUSSION

The aim of this study was to identify a pattern of features, and then apply an existing RE language for the modeling of the requirements implemented by those features. Indeed, as far as we know no literature addresses this problem, and it was interesting to find out whether the development of some RE language specific to the design of an OSN is necessary. We applied the i* language to the set of requirements discovered earlier. The study shows that this language, and more specifically the Strategic Rationale Model, allows the modeling of the requirements.

The pattern-based RE approach can be seen as a type of market-driven RE approach (Regnell & Brinkkemper, 2005). Indeed, in a market-driven RE (MDRE) approach, a software product is developed for an open market with many customers, and not for a specific customer. In MDRE, requirements elicitation focuses on a combination of market analysis and innovation of new requirements. In MDRE, time and effort are devoted to prioritization, over negotiation and conflict resolution. Here, we identified common features from popular OSNs. In (Bouraga et al., 2015), we focused on the prioritization of these features. The aim of the end product is to attract as many users as possible.

From Section 4, we can see that the main difference between OSNs is not in terms of the presence or absence of the features; but the main difference is rather in the form taken by the features. For instance, for the content features, all of the OSNs except for Pinterest allow their users to post some text. The distinction is made on the form of the text: for Facebook and MySpace it is a status; for Twitter, it is a tweet; etc. As far as the profile features are concerned, Facebook, LinkedIn, and MySpace are the three OSNs that enable their users to share the most information. Half of the OSNs (Facebook, Flickr, LinkedIn, and YouTube) allow for both types of relationships between users; MySpace allows only the bidirectional type; and the rest of them (Pinterest, Tumblr, and Twitter) enable the uni-directional relationships. In terms of content, most of the OSNs provide all the features listed here, in one form or another. Only Pinterest does not provide the text feature. As far as the privacy is concerned, five OSNs out of the 8 studied here implement the three degrees of privacy. MySpace and Twitter do not let their users to have private posts; and Tumblr does not allow for semi-private posts. The goal feature demonstrates that four OSNs (Flickr, Pinterest, Tumblr, and YouTube) are of the knowledge-sharing type; while Facebook, LinkedIn, and Twitter are network-oriented. We can identify MySpace as being of both types. All OSNs generate recommendations about other users; all OSNs but Facebook and Twitter produce recommendations about content; and only three OSNs (Facebook, LinkedIn, and Twitter) generate recommendations about a Business or a Celebrity. Finally, all OSNs allow a connection with another OSN to sign in and to share content; except for Flickr, LinkedIn, and Tumblr. The latter only allow their users to share content simultaneously on different OSNs, but they do not allow their users to sign in with another OSN.

After discovering the set of features, we sought to model them using the i* language. The i* language allows the representation of the actors' goals, tasks, and resources. It also enables the representation of the dependencies between actors: who depends on whom to do what? The i* language is expressive enough to allow the representation of every specific requirement identified here. We can also represent the precise relationships between all the actors involved in the situation considered here. We can specify that the OSN can display all kinds of information (for instance, the profile information and the content), but the OSN depends on the user to provide the information. We can also notice that in order to link two users together, the OSN depends on both users to have an account and provide several types of resources. This observation leads us to claim that i* is appropriate for the representation of the OSN requirements. Moreover, i* allows us to model the non-functional requirements using the softgoal concept. We consider that the user has the following softgoals: she wants quality recommendations, she wants to enjoy her experience on the OSN, and she wants her privacy to be respected. Some of the user's tasks contribute positively to the softgoal. Namely, the task "Share personal information" contributes positively to the softgoal "Quality recommendations", indeed, the more personal information the user shares, the more the OSN knows her, the more the OSN is able to generate relevant recommendations. The tasks "Get recommendations" and "Connect with other OSNs" contribute positively to the softgoal "Enjoy experience"; and the task "Manage privacy settings" contributes positively to the softgoal "Privacy respected". As far as the OSN is concerned, we identify - at least - two softgoals: "Satisfy the user" and "Make money".

We believe that these patterns have important practical implications for the design of new OSNs. They reflect best practices we have seen so far. Using the model we proposed here should help jumpstart the development of new OSNs. It can be used as a baseline or starting point in the RE step of OSN engineering, in order to, for instance, stimulate ideas during elicitation; to better understand, and explicitly state the elements that should be taken into account in RE for OSNs; to help avoid the pitfall of implicit requirements; to help check completeness of a requirements model for OSN; and so on. The proposed patterns are generic enough so that they can be adapted to a particular environment. Yet, we believe we were also complete and managed to represent key features.

We will now apply our pattern to Twitter. We model the requirements that Twitter has to satisfy, in order to show an application of our model. Firstly, we replaced the generic terms in the pattern by the specific terms used by Twitter. For instance, instead of using the term post, we use the terms Tweet, Retweet, and Favorite. Also, on Twitter, only uni-directional relationships are considered. Hence, we removed the part about bi-directional relationships. The resulting models are represented in Figures 4 and 5.

CONCLUSION

In this paper, we offered an analysis of recurrent features in relation to several OSNs, namely Facebook, Flickr, LinkedIn, MySpace, Pinterest, Tumblr, Twitter and YouTube; which is summarized in Tables 2 to 8.

Figure 4. Strategic Rationale Model for Twitter - Part 1

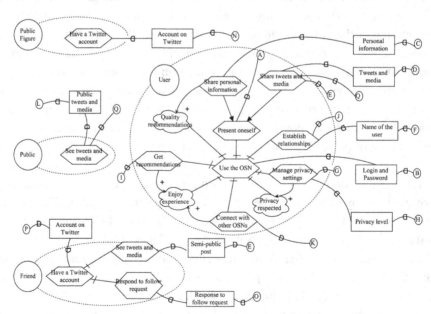

Figure 5. Strategic Rationale Model for Twitter - Part 2

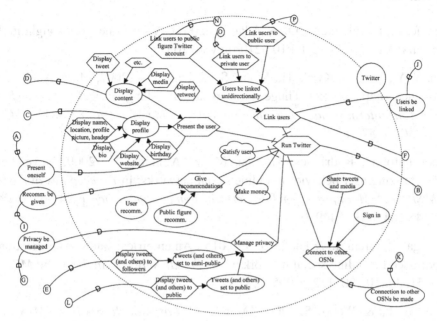

We also represented the pattern of features using an existing requirements engineering language, namely i*. We found out that this model is appropriate to represent the various requirements, and it also allows to model the relationships between the various actors. We discussed our results and applied our pattern to Twitter.

We believe that this study adds to existing research in several ways. Firstly, as far as we know, this is the first research to address OSNs in terms of: (i) identifying requirements implemented by OSN features, and (ii) modeling these requirements using i*. And secondly, we believe it can have implications for the design of future OSNs; more specifically for the RE phase. Indeed, when designing an OSN, we now have an available starting point, which synthesizes the many features that may be needed to satisfy the requirements and expectations of OSN users and other stakeholders.

REFERENCES

Adamic, L., Buyukkokten, O., & Adar, E. (2003). A social network caught in the web. *First Monday*, *8*(6). doi:10.5210/fm.v8i6.1057

Ahn, Y. Y., Han, S., Kwak, H., Moon, S., & Jeong, H. (2007, May). Analysis of topological characteristics of huge online social networking services. In *Proceedings of the 16th international conference on World Wide Web* (pp. 835-844). ACM. 10.1145/1242572.1242685

Benevenuto, F., Rodrigues, T., Cha, M., & Almeida, V. (2009, November). Characterizing user behavior in online social networks. In *Proceedings of the 9th ACM SIGCOMM conference on Internet measurement conference* (pp. 49-62). ACM. 10.1145/1644893.1644900

Bouraga, S., Jureta, I., & Faulkner, S. (2015). An empirical study of notifications' importance for online social network users. *Social Network Analysis and Mining*, *5*(1), 51. doi:10.100713278-015-0293-x

Caverlee, J., & Webb, S. (2008, March). *A Large-Scale Study of MySpace: Observations and Implications for Online Social Networks*. ICWSM.

Centola, D. (2010). The spread of behavior in an online social network experiment. *Science*, *329*(5996), 1194-1197.

Chao, J. T., Parker, K. R., & Fontana, A. (2011). Developing an interactive social media based learning environment. *Issues in Informing Science and Information Technology*, *8*, 323–334. doi:10.28945/1421

Ducheneaut, N., Yee, N., Nickell, E., & Moore, R. J. (2006, April). Alone together?: exploring the social dynamics of massively multiplayer online games. In *Proceedings of the SIGCHI conference on Human Factors in computing systems* (pp. 407-416). ACM. 10.1145/1124772.1124834

Dudenhoffer, C. (2012). Pin it!: Pinterest as a library marketing and information literacy tool. *College & Research Libraries News*, *73*(6), 328–332. doi:10.5860/crln.73.6.8775

Dwyer, C., Hiltz, S. R., & Passerini, K. (2007, August). Trust and Privacy Concern Within Social Networking Sites: A Comparison of Facebook and MySpace. In AMCIS (p. 339). Academic Press.

Ellison, N. B. (2007). Social network sites: Definition, history, and scholarship. *Journal of Computer-Mediated Communication*, *13*(1), 210–230. doi:10.1111/j.1083-6101.2007.00393.x

Facebook. (n.d.). Retrieved from www.facebook.com

Fu, F., Liu, L., & Wang, L. (2008). Empirical analysis of online social networks in the age of Web 2.0. *Physica A, 387*(2), 675–684. doi:10.1016/j.physa.2007.10.006

Ghafoor, F., & Niazi, M. A. (2016). Using social network analysis of human aspects for online social network software: A design methodology. *Complex Adaptive Systems Modeling, 4*(1), 14. doi:10.118640294-016-0024-9

Gilbert, E., Bakhshi, S., Chang, S., & Terveen, L. (2013, April). I need to try this?: a statistical overview of pinterest. In *Proceedings of the SIGCHI conference on human factors in computing systems*(pp. 2427-2436). ACM. 10.1145/2470654.2481336

Gill, P., Arlitt, M., Li, Z., & Mahanti, A. (2007, October). Youtube traffic characterization: a view from the edge. In *Proceedings of the 7th ACM SIGCOMM conference on Internet measurement* (pp. 15-28). ACM. 10.1145/1298306.1298310

Golbeck, J. (2009). Trust and nuanced profile similarity in online social networks. *ACM Transactions on the Web, 3*(4), 12. doi:10.1145/1594173.1594174

Golder, S. A., Wilkinson, D. M., & Huberman, B. A. (2007). Rhythms of social interaction: Messaging within a massive online network. *Communities and technologies 2007*, 41-66.

Gopsill, J. A., McAlpine, H. C., & Hicks, B. J. (2013). A social media framework to support engineering design communication. *Advanced Engineering Informatics, 27*(4), 580–597. doi:10.1016/j.aei.2013.07.002

Guha, S., Tang, K., & Francis, P. (2008, August). NOYB: Privacy in online social networks. In *Proceedings of the first workshop on Online social networks* (pp. 49-54). ACM. 10.1145/1397735.1397747

Guo, L., Tan, E., Chen, S., Zhang, X., & Zhao, Y. E. (2009, June). Analyzing patterns of user content generation in online social networks. In *Proceedings of the 15th ACM SIGKDD international conference on Knowledge discovery and data mining* (pp. 369-378). ACM. 10.1145/1557019.1557064

Gurses, S., Rizk, R., & Gunther, O. (2008). Privacy design in online social networks: Learning from privacy breaches and community feedback. *ICIS 2008 Proceedings*, 90.

Heer, J., & Boyd, D. (2005, October). Vizster: Visualizing online social networks. In *Information Visualization, 2005. INFOVIS 2005. IEEE Symposium on* (pp. 32-39). IEEE.

Java, A., Song, X., Finin, T., & Tseng, B. (2007, August). Why we twitter: understanding microblogging usage and communities. In *Proceedings of the 9th WebKDD and 1st SNA-KDD 2007 workshop on Web mining and social network analysis* (pp. 56-65). ACM. 10.1145/1348549.1348556

Kaplan, A. M., & Haenlein, M. (2010). Users of the world, unite! The challenges and opportunities of Social Media. *Business Horizons, 53*(1), 59–68. doi:10.1016/j.bushor.2009.09.003

Karampelas, P. (2012). *Techniques and tools for designing an online social network platform.* Springer.

Kietzmann, J. H., Hermkens, K., McCarthy, I. P., & Silvestre, B. S. (2011). Social media? Get serious! Understanding the functional building blocks of social media. *Business Horizons, 54*(3), 241–251. doi:10.1016/j.bushor.2011.01.005

Kumar, R., Novak, J., & Tomkins, A. (2010). Structure and evolution of online social networks. In *Link mining: models, algorithms, and applications* (pp. 337–357). Springer New York. doi:10.1007/978-1-4419-6515-8_13

Kwak, H., Lee, C., Park, H., & Moon, S. (2010, April). What is Twitter, a social network or a news media? In *Proceedings of the 19th international conference on World wide web* (pp. 591-600). ACM. 10.1145/1772690.1772751

Lai, R. K., Tang, J. C., Wong, A. K., & Lei, P. I. (2010). Design and implementation of an online social network with face recognition. *Journal of Advances in Information Technology, 1*(1), 38–42. doi:10.4304/jait.1.1.38-42

Lampe, C., Ellison, N., & Steinfield, C. (2006, November). A Face (book) in the crowd: Social searching vs. social browsing. In *Proceedings of the 2006 20th anniversary conference on Computer supported cooperative work* (pp. 167-170). ACM. 10.1145/1180875.1180901

Lampe, C. A., Ellison, N., & Steinfield, C. (2007, April). A familiar face (book): profile elements as signals in an online social network. In *Proceedings of the SIGCHI conference on Human factors in computing systems* (pp. 435-444). ACM. 10.1145/1240624.1240695

Lange, P. G. (2007). Publicly private and privately public: Social networking on YouTube. *Journal of Computer-Mediated Communication, 13*(1), 361–380. doi:10.1111/j.1083-6101.2007.00400.x

Leskovec, J., Huttenlocher, D., & Kleinberg, J. (2010, April). Predicting positive and negative links in online social networks. In *Proceedings of the 19th international conference on World wide web* (pp. 641-650). ACM. 10.1145/1772690.1772756

Liu, H. (2007). Social network profiles as taste performances. *Journal of Computer-Mediated Communication, 13*(1), 252–275. doi:10.1111/j.1083-6101.2007.00395.x

Livingstone, S. (2008). Taking risky opportunities in youthful content creation: Teenagers' use of social networking sites for intimacy, privacy and self-expression. *New Media & Society, 10*(3), 393–411. doi:10.1177/1461444808089415

Madejski, M., Johnson, M. L., & Bellovin, S. M. (2011). *The failure of online social network privacy settings*. Academic Press.

Magoutis, K., Papoulas, C., Papaioannou, A., Karniavoura, F., Akestoridis, D. G., Parotsidis, N., ... Stephanidis, C. (2015). Design and implementation of a social networking platform for cloud deployment specialists. *Journal of Internet Services and Applications, 6*(1), 19. doi:10.118613174-015-0033-5

Marlow, C., Naaman, M., Boyd, D., & Davis, M. (2006, August). HT06, tagging paper, taxonomy, Flickr, academic article, to read. In *Proceedings of the seventeenth conference on Hypertext and hypermedia* (pp. 31-40). ACM. 10.1145/1149941.1149949

Mislove, A., Koppula, H. S., Gummadi, K. P., Druschel, P., & Bhattacharjee, B. (2008, August). Growth of the flickr social network. In *Proceedings of the first workshop on Online social networks* (pp. 25-30). ACM. 10.1145/1397735.1397742

Mislove, A., Marcon, M., Gummadi, K. P., Druschel, P., & Bhattacharjee, B. (2007, October). Measurement and analysis of online social networks. In *Proceedings of the 7th ACM SIGCOMM conference on Internet measurement* (pp. 29-42). ACM. 10.1145/1298306.1298311

Mislove, A., Viswanath, B., Gummadi, K. P., & Druschel, P. (2010, February). You are who you know: inferring user profiles in online social networks. In *Proceedings of the third ACM international conference on Web search and data mining* (pp. 251-260). ACM. 10.1145/1718487.1718519

Nosko, A., Wood, E., & Molema, S. (2010). All about me: Disclosure in online social networking profiles: The case of FACEBOOK. *Computers in Human Behavior, 26*(3), 406–418. doi:10.1016/j.chb.2009.11.012

Papacharissi, Z. (2009). The virtual geographies of social networks: A comparative analysis of Facebook, LinkedIn and ASmallWorld. *New Media & Society, 11*(1-2), 199–220. doi:10.1177/1461444808099577

Park, N., Kee, K. F., & Valenzuela, S. (2009). Being immersed in social networking environment: Facebook groups, uses and gratifications, and social outcomes. *Cyberpsychology & Behavior, 12*(6), 729–733. doi:10.1089/cpb.2009.0003 PMID:19619037

Pempek, T. A., Yermolayeva, Y. A., & Calvert, S. L. (2009). College students' social networking experiences on Facebook. *Journal of Applied Developmental Psychology, 30*(3), 227–238. doi:10.1016/j.appdev.2008.12.010

Pereira, R., Baranauskas, M. C. C., & da Silva, S. R. P. (2010, June). Social software building blocks: revisiting the honeycomb framework. In *Information Society (i-Society), 2010 International Conference on* (pp. 253-258). IEEE.

Plotnick, L., White, C., & Plummer, M. M. (2009). The design of an online social network site for emergency management: A one stop shop. *AMCIS 2009 Proceedings*, 420.

Raacke, J., & Bonds-Raacke, J. (2008). MySpace and Facebook: Applying the uses and gratifications theory to exploring friend-networking sites. *Cyberpsychology & Behavior, 11*(2), 169–174. doi:10.1089/cpb.2007.0056 PMID:18422409

Regnell, B., & Brinkkemper, S. (2005). *13 Market-Driven Requirements Engineering for Software Products*. Academic Press.

Schneider, F., Feldmann, A., Krishnamurthy, B., & Willinger, W. (2009, November). Understanding online social network usage from a network perspective. In *Proceedings of the 9th ACM SIGCOMM conference on Internet measurement conference* (pp. 35-48). ACM. 10.1145/1644893.1644899

Sheth, S., Kaiser, G., & Maalej, W. (2014, May). Us and them: a study of privacy requirements across North America, Asia, and Europe. In *Proceedings of the 36th International Conference on Software Engineering* (pp. 859-870). ACM. 10.1145/2568225.2568244

Skeels, M. M., & Grudin, J. (2009, May). When social networks cross boundaries: a case study of workplace use of facebook and linkedin. In *Proceedings of the ACM 2009 international conference on Supporting group work* (pp. 95-104). ACM. 10.1145/1531674.1531689

Strater, K., & Lipford, H. R. (2008, September). Strategies and struggles with privacy in an online social networking community. *Interaction, 1*, 111–119.

Subrahmanyam, K., Reich, S. M., Waechter, N., & Espinoza, G. (2008). Online and offline social networks: Use of social networking sites by emerging adults. *Journal of Applied Developmental Psychology, 29*(6), 420–433. doi:10.1016/j.appdev.2008.07.003

Valenzuela, S., Park, N., & Kee, K. F. (2009). Is There Social Capital in a Social Network Site?: Facebook Use and College Students' Life Satisfaction, Trust, and Participation1. *Journal of Computer-Mediated Communication, 14*(4), 875–901. doi:10.1111/j.1083-6101.2009.01474.x

Van Der Sype, Y. S., & Maalej, W. (2014, August). On lawful disclosure of personal user data: What should app developers do? In *Requirements Engineering and Law (RELAW), 2014 IEEE 7th International Workshop on* (pp. 25-34). IEEE.

Van Lamsweerde, A. (2007). *Requirements engineering.* John Wiley & Sons.

Van Lamsweerde, A. (2008, November). Requirements engineering: from craft to discipline. In *Proceedings of the 16th ACM SIGSOFT International Symposium on Foundations of software engineering* (pp. 238-249). ACM. 10.1145/1453101.1453133

Viswanath, B., Mislove, A., Cha, M., & Gummadi, K. P. (2009, August). On the evolution of user interaction in facebook. In *Proceedings of the 2nd ACM workshop on Online social networks* (pp. 37-42). ACM. 10.1145/1592665.1592675

Weiss, J. B., Berner, E. S., Johnson, K. B., Giuse, D. A., Murphy, B. A., & Lorenzi, N. M. (2013). Recommendations for the design, implementation and evaluation of social support in online communities, networks, and groups. *Journal of Biomedical Informatics, 46*(6), 970–976. doi:10.1016/j.jbi.2013.04.004 PMID:23583424

Xiang, R., Neville, J., & Rogati, M. (2010, April). Modeling relationship strength in online social networks. In *Proceedings of the 19th international conference on World wide web* (pp. 981-990). ACM. 10.1145/1772690.1772790

Yu, E. S. (1997, January). Towards modelling and reasoning support for early-phase requirements engineering. In *Requirements Engineering, 1997., Proceedings of the Third IEEE International Symposium on* (pp. 226-235). IEEE. 10.1109/ISRE.1997.566873

Zhang, W., Nie, L., Jiang, H., Chen, Z., & Liu, J. (2014). Developer social networks in software engineering: Construction, analysis, and applications. *Science China. Information Sciences, 57*(12), 1–23.

Chapter 3

Advances in E-Pedagogy for Online Instruction:
Proven Learning Analytics Rasch Item Response Theory

Elspeth McKay
RMIT University, Australia

Allaa Barefah
RMIT University, Australia

Marlina Mohamad
*Universiti Tun Hussein Onn Malaysia,
Malaysia*

Mahmoud N. Bakkar
Holmes Institute, Australia

ABSTRACT

Many people are finding it relatively easy to engage with courseware development and simply upload it to the internet. The trouble with this approach is there are no quality controls to ensure the impending instructional strategies are designed well. This chapter presents a set of research projects that incrementally focus on instructional strategies as they apply in today's information communications technology (ICT) tools. They commence with a simple investigation into online courseware matching cognitive preferences. Next, a project extends this principle and takes the research to a government-training context to concentrate on dealing with a broader range of stakeholders to customize training for a user-centered environment. Then the authors present a project designed for mobile healthcare training on an iPad. The final research project synthesizes the current instructional systems design (ISD) thinking to promote a prescriptive information systems (IS)-design model for educational courseware, offering it to the ISD research community as an extension to the ADDIE training development model.

DOI: 10.4018/978-1-5225-5715-9.ch003

INTRODUCTION

Understanding the effects of instructional strategies on learning outcomes has been a continual topic of interest among educational technology researchers (McKay, 2018); (Merrill, Barclay, & Van-Schaak, 2016);(Cooper-Smith & McKay, 2015);(Minor, 2014);(Parker, 2004). To this end, the Mohamad and McKay (2014) joined the research community to examine various aspects of a web-mediated instructional system (WMIS) in the acquisition of programming skills to reveal that novice-learners performed better with text-plus-textual metaphors than text-plus-graphical metaphors regardless of their cognitive preferences (Allwood, Traum, & Jokinen, 2000).

The main purpose of this paper is to provide an extension of the Mohamad and McKay (2014) paper and present the research developments since. Before presenting three new contributions, to set the context there will be an overview of the initial 2014 paper; this work reported on the Mohamad (2012) thesis, which examined the interactive effects of online instruction and cognitive style preferences for learning computer programming. Then, the first of the new research projects describes how the researchers prepared for their funded experimental research project. This involved an investigation of the interactive effect of instructional strategies (face-to-face; a blend of face-to-face and computerised instruction; and computerised only facilitation), and cognitive instructional preferences for gaining introductory ethics knowledge and skills in the public sector for government trainees (McKay & Izard, 2014). The second new research project describes a part of the Bakkar (2016) thesis to concentrate on the same instructional design (ID) process used by Mohamad (2012) to highlight the importance of conducting a fine-grained approach to planning courseware development for a mobile device. Last but not the least important is the work of Allaa Barefah as she prepares to complete her thesis in the coming months. This work is a synthesis of the earlier research projects to advocate for a prescriptive IS-design model (McKay, 2018) to enhance user-centred instruction.

The Mohamad (2012) thesis used the cognitive styles analysis (CSA) developed by Riding and Cheema (1991) to determine participants' preferred cognitive instructional mode (verbaliser-imager (VI) dimension that depicts the way people represent information during thinking and the wholist-analytic (WA) dimension which explains the mode of processing information (Riding, 2005). The participants were given the CSA to determine their cognitive styles ratio (VI:WA) that was used to allocate the instructional treatment (text-plus-textual metaphor (T1) and the text-plus-graphical metaphor (T2) (see Figure-1). Initially there were 399-participants who took part in this experiment. They were second year undergraduate students enrolled in a Malaysian university for a Bachelor of Civil Engineering course (Mohamad, 2012). However, the data from 47-participants were discarded from the analysis due to not sitting for the CSA test or they had incomplete answers to the pre/post questionnaires. Therefore, only 352-participants were included in the data analysis as they did complete the whole experiment (Mohamad & McKay, 2014).

When considering the integrated cognitive preference (ICP) of the participants' CSA ratio (wholist-verbaliser (WV); wholist-imager (WI); analytic-verbaliser (AV); analytic-imager (AI), the Mohamad (2012) study found that analytic-verbalisers performed considerably better with the text-plus-textual metaphors (T1) compared to their counterparts given the text-plus-graphical metaphors (T2), see Figure-2. This was probably so because the analytic-verbalisers may have had good verbal memories that more easily retain the information, particularly when presented in a textual (or verbal) format (Riding, 2005).

Furthermore, it was thought by Mohamad (2012) that when presented information in a structured verbal format the analytic-verbalisers no longer found pictures or graphics very helpful. However, by adding some 'visual signals' or 'cues' may assist learning from textual instruction (Mautone & Mayer, 2001). Such verbal informational signals: paragraph headings; bolding and italicizing of important words; signals phrases; pointer words and topical overviews written as content preview (Clark, Nguyen, & Sweller, 2006)p.79;(Mayer, 2001). For instance, showing how signals in the tabular form, added to the instructional materials in the text-plus-textual

Figure 1. CSA used to allocate instructional treatments (T1 and T2)
Source: Mohamad and McKay (2014 p.25)

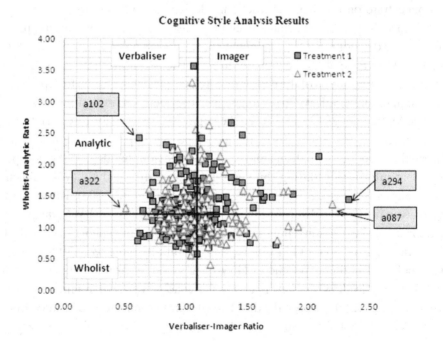

Figure 2. Interactive effects of ICP and online instructional formats
Source: Mohamad and McKay (2014 p.30)

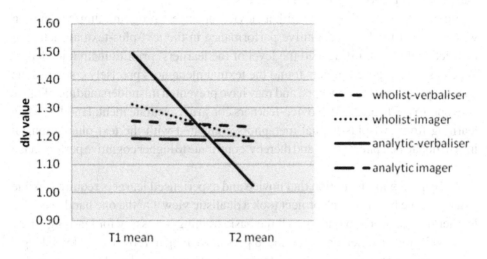

instructional metaphor format for comparing three control structures: 'while', 'do.. while' and 'for.' This effect might help the analytic-verbalisers to retain information presented to them longer, because the purpose of a conceptual 'signal' is to draw attention to essential information in the lesson, yet not to add any new information, according to Mayer (2009).

In addition Mohamad (2012) found the cognitive performance of novice wholist-imagers and analytic-imagers were almost similar with text-plus-textual metaphor and text-plus-graphical metaphor instructional formats. In other words, the cognitive performance of the imagers was not affected by the instructional strategies. Presenting programming concepts in text-plus-graphical metaphor (or multimedia) format did not assist the imagers in this study because they performed best when the instructional materials were visualised in mental pictures that did not contain any acoustically complex and unfamiliar terms (such as programming language) (Riding, 2005). This finding was contra to what has been found by previous research (Riding & Douglas, 1993) which discovered that imagers learned better when the materials on motor cars braking systems were presented in a text-plus-illustration format, rather than with text alone. McKay (2000) also found that imagers were superior to verbalisers with text plus graphical metaphors. Similarly, when tested on college and university students, Smith and Woody (2000) observed that visualisers learned better than verbalisers when given learning materials in a multimedia format. Further evidence, revealed that the learning performance of wholist- and analytic-imagers was enhanced with a visual metaphor interface (Lee, 2007). Alwi and McKay (2010) also discovered improvement in learning outcomes when instructional materials match with students' cognitive preferences.

Finally, when dealing with either novice or expert programming students, it was clear that the better cognitive performance in the text-plus-textual metaphor instructional format influenced the level of the learners' prior domain knowledge. Whereas novice-programmers found the textual metaphors precisely conveyed the abstract programming concepts, and may have prevented misunderstanding of those concepts (Yu-Chen, 2007). Novice-learners construct accurate mental models when learning from text-plus-textual metaphors compared with the text-plus-graphical instructional metaphor format and thereby, contribute to higher cognitive performance outcomes.

In keeping with the notion that novice and experienced learners require flexible strategies, the next research project took a dualistic view. On the one hand, a novice learner/trainee needs to undergo all the basic training necessary for learning a new skill; while on the other, an experienced person wishing to refresh their knowledge/skills needs to choose the order of their instructional modules. This next section describes how the researchers prepared for their funded experimental research project. The research involved an investigation of the interactive effect of instructional strategies (face-to-face; a blend of face-to-face and computerised instruction; and computerised only facilitation), and cognitive instructional preferences for gaining introductory ethics knowledge and skills in the public sector for government trainees (McKay & Izard, 2014).

CUSTOMISING FOR USER-CENTERED PARTICIPATION

The government sector relies on continual employee reskilling through cost effective eLearning programmes that rely upon information communications technology (ICT) tools to enhance work-place training that assure predictable outcomes (McKay & Vilela, 2011). It has been found that the most desirable approach is to personalise the opportunities to increase an employee's knowledge development through flexible online learning; while improved IT governance serves to motivate disinterested trainees and energise frustrated management (McKay, Axmann, Banjanin, & Howat, 2007). Multi-disciplined specialists are required to resolve the factional dilemmas of corporate IT resource ownership. The timeliness of this project highlighted desirable change management issues to improve efficiencies and effectiveness of existing IT training resources. Maintaining well skilled and knowledgeable employees is key to sustaining our competitive advantage through smarter information use of digital technologies. This project served to distribute the design techniques to enable novice courseware developers design and build their own online learning programmes drawing upon sound instructional design principles (Merrill, Barclay, & Van-Schaak, 2008). We now know that access to our individual virtual learning space is critical (Boddington & Boys, 2011). Therefore, it was apparent that courseware design techniques must draw on Web 2.0 technologies to empower trainee/learner with the global reach of an individual's online access to adaptable eLearning tools that provide the unyielding intellectual thirst for new learning spaces for the next decade (McKay, 2008).

The primary aim of this funded project was to investigate how to distribute the mass of corporate knowledge contained in large government organisations through smart information use. The primary motivation for this work was to address a national government's priority research goal: 'improved data management for existing and new business applications and creative applications for digital technologies.' The researchers evaluated the effectiveness of ICT tools as an intelligent AGENT-based adaptive training aids; by direct contrast with existing methods that were currently being used by several government agencies (McKay & Izard, 2015b). The project team has already disseminated their understandings and insights on the hurdles they overcame to shift the negative perceptions towards designing customised eLearning solutions (McKay & Izard, 2016; McKay & Izard, 2015); (McKay & Izard, 2015a); (McKay & Izard, 2015; McKay & Izard, 2014).

The overall research design was a quasi-experimental 3x3 factorial design, involving three independent variables: instructional format; learning preference; and prior domain knowledge. There were three levels for each variable. The three instructional formats under investigation were: (1) no computer-based material comprising traditional face-to-face instructor led classroom approach; (2) a blended approach involving both traditional face-to-face and eTraining tools; and (3) eLearning as the central computerised instructional strategy. Learning preference was categorised as either: (1) verbal; (2) intermediate; and (3) imagery in a paper-based questionnaire that determined participants' learning preference (Blazhenkova & Kozhevnikov, 2009). It was important to identify each participant's learning preference before randomly assigned to one of three instructional treatment groups. The prior domain knowledge variable identified: (1) novice, (2) intermediate, and (3) experienced; operationalised when a screening test for each participant conducted before the training session. Novice learners have been defined as having little relevant prior knowledge in their memory (Bagley & Heltne, 2003). As a result, their cognitive processes are different from an employee with greater background experience. The results concurred with Bagley and Heltne (2003) showing that novice trainees benefitted from different instructional methods than more experienced trainees.

The participants met in a lecture theatre; those that consented to their participation were registered and given a Research Code (R-Code) to anonymize the data. This R-Code was written on both their Object-Spatial Imagery and Verbal Questionnaire (OSIVQ) (Blazhenkova & Kozhevnikov, 2009), and their pre-test questionnaire. This instrument was used to distinguish: the object imagers who prefer to construct vivid, concrete and detailed images of individual objects; the spatial imagers who use imagery to schematically represent spatial relations among objects and to perform complex spatial transformations (like scientists); and the verbalisers who prefer to use verbal-analytical tools to solve cognitive tasks (like philosophers and linguists). Upon completion, the OSIVQ, pre-test questionnaire and the preferred training mode handout (Figure 3) collected for data analysis.

Figure 4 shows how the researchers managed to keep the integrity of the 3 x 3 design with the following schedule.

Figure 5 shows how the instructional treatment allocation generated through the P-Code registration listing. Participants were advised treatment group before the first tea/coffee break. Note that T1 participants remained seated for Module-1.

During the digital treatments, the data was captured online through the (SCORM compliant) personalised Avatar knowledge navigation tool (Figure-6) (McKay & Izard, 2013). SCORM known as a 'sharable content object reference model,' is a collection of specifications that enable interoperability, accessibility and reusability of Web-based learning content (McKay & Izard, 2014). Figure-7 shows how the training

Figure 3. Preferred instructional mode
Source: Adapted from McKay and Izard (2014)

Preferred Training Mode –
enter the letter(s) in the box:
F   = face-to-face (classroom)
ED = e-learning at the work desk
EE = e-learning elsewhere at work or other venue
EH = e-learning at home
C   = combined face-to-face and e-learning at work or other venue
P   = paper-based correspondence

Experience with e-learning – enter number in diamond:
3 = less than 4 hours
6 = 4 to 6 hours
9 = more than 6 hours

Should in-service learning for work purposes by any mode
be considered part of one's workload? –
enter in letter in the circle:
Y = yes, always
M = maybe, depending on circumstances
N = no, never

Figure 4. Research schedule
Source: McKay and Izard (2014, p.3)

		9.00-9.15	9.15	9.30AM	9.30-9.45	9.45-12.30 Ethics Course				12.30-1 PM
						Module 1	Module 2	Module 3	Module	

e-LEARNING EXPERIMENT - RESEARCH SCHEDULE - ROOM REQUIREMENTS

RESEARCH GROUPS

Face-2-Face PARTICIPANTS (25) — PRE-TEST & FACILITATOR'S — T/C Break — Classroom participation -------- CLASSROOM MUST HOLD 25 PARTICIPANTS

Blended PARTICIPANTS (25) — Module 1 — T/C Break — -------- Computerised -------- - LAB MUST HOLD 25 PAARTICIPANTS

Computerised PARTICIPANTS (25) — T/C BREAK — -------- Computerised -------- -- LAB MUST HOLD 25 PARTICIPANTS

Register & Short Exercise on Paper — BRIEFING

POST-TEST & OTHER O

courseware afforded user-centred choice of instructional delivery. The beginner (novice level) course enabled a step-by-step approach to the skills development training; whereas the experienced (refresher level) course enabled the trainee to choose their own skills building module order.

This research project involved a pilot study to validate the performance measurement instrumentation (McKay & Izard, 2015) and the research schedule (see Figure-4). In contrast with the pilot study, which was conducted with 20-government training sector participants, the main experiment involved 45-non-government learning institution participants, from an industry training programme.

Evaluation of the participants' learning outcomes employed test-item response modelling with the 'QUEST interactive analysis system' (Wu & Adams, 2007);(Adams & Khoo, 1996). This test measurement application allows for improved analyses of an individual's performance relative to other participants and relative to the test-item difficulty of introductory ethics knowledge levels (Bond & Fox, 2015);(McKay, 2000). See an example of this in Figure-8, where each participant or 'case' is depicted by an 'x,' and test-items are shown on the right-side of the map. The Rasch Item Response Theory (IRT) estimates the probability of an individual making a certain response to a test-item. The pre- and post-test results were analysed with a test-item matrix that had each individual's responses for every test-item recorded. Common test-items

Figure 5. Instructional treatment allocation
Source: McKay and Izard (2014, p.3)

P-Code:	T-Colour	T-Group
2001	White	T1
2002	green	T-2
2003	red	T-3
2004	white	T-1
2005	green	T-2
2006	red	T-3

*For a more accurate representation see the electronic version.

Figure 6. System AGENT
Source: McKay and Izard (2014, pp.1-2)

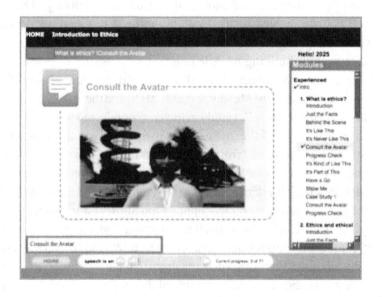

Figure 7. Courseware experience options
Source: McKay and Izard (2014, pp.1-2)

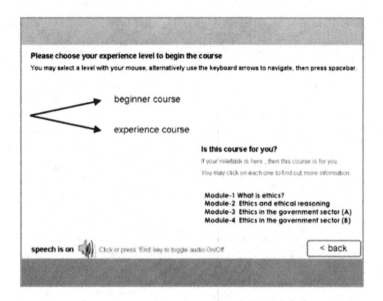

(identically worded questions) were 'anchored' so that scale scores on the pre-test were comparable with scale scores on the post-test. The difference between pre-test and post-test scaled scores indicated whether learning occurred, whether no learning occurred, or whether the instructional strategy resulted in reduced achievement.

The annotated the QUEST variable map (Figure 8) revealed that participants X11, X12, and X18 achieved the same level of knowledge at a scale score of -0.80. The colour of the numeral indicates the treatment group (black is face-to-face T1, green is blended T2, and red is computerised T3). Participants X07, X15, and X30 achieved the same level of knowledge at a scale score of 1.10 close to the average difficulty of the Pre-Test (0.0), and this level was higher than participants X11, X12, and X18. This annotated variable map also showed that scoring a 1 on item-5 is more difficult than scoring a 1 on item-12. Getting item-19 correct is easier than scoring 2 or more on either item-16 or item-18. Scoring 4 on item-17 is easier that scoring 2 or more on either item-16 or item-18. The probability basis of the model illustrated by the position of participants X05, X08, and X32 at an achievement level of 1.56. These participants are more likely than not to score 1 on items 15, 16, and 18 and 3 on item 17 (McKay & Izard, 2014 p.5).

Figure 8. Annotated QUEST post-test variable map
Source: McKay and Izard (2014, p.5)

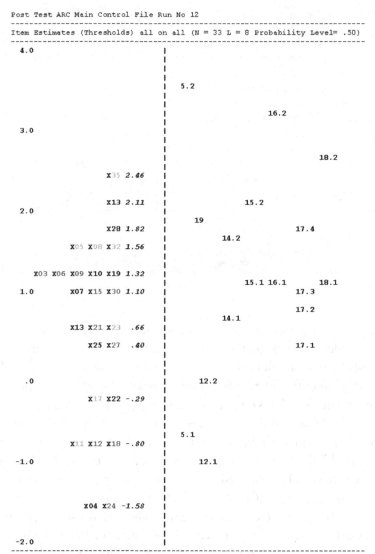

Post Test ARC Main Control File Run No 12

Item Estimates (Thresholds) all on all (N = 33 L = 8 Probability Level= .50)

```
        4.0              |
                         |
                         |
                         |
                         |    5.2
                         |
                         |                    16.2
        3.0              |
                         |
                         |                           18.2
                         |
              X35  2.46   |
                         |
              X13  2.11   |           15.2
        2.0              |
                         |   19
              X28  1.82   |                    17.4
     X05 X08 X32  1.56    |        14.2
                         |
  X03 X06 X09 X10 X19  1.32 |
                         |        15.1 16.1      18.1
        1.0     X07 X15 X30  1.10 |              17.3
                         |                       17.2
                         |        14.1
        X13 X21 X23  .66  |
           X25 X27  .40   |                      17.1
                         |
         .0              |    12.2
                         |
           X17 X22 -.29   |
                         |
                         |
                         |   5.1
        X11 X12 X18 -.80  |
       -1.0              |    12.1
                         |
                         |
                         |
        X04 X24 -1.58     |
                         |
       -2.0              |
```

Each X represents 1 student; the associated numeral identifies the student.
High achievement status is at the top and low at the bottom. Difficult items
are shown near the top, easy near the bottom.
==

The project described above facilitates cost effective eLearning practice using advanced ICT tools to enhance work-place training with assured predictable outcomes. The most desirable training approach is to personalise an employee's knowledge development through flexible online learning. Improved IT governance serves to motivate disinterested trainees and energise frustrated management. However, within this digital training realm multi-disciplined specialists are required to resolve the factional dilemmas of corporate IT resource ownership. The timeliness of this project will highlight desirable change management issues to improve efficiencies and effectiveness of existing IT training resources.

This project kept the focus on ID and instructional systems design (ISD) adding a government training perspective of adult learning in a more public forum than higher education (HE) per se. The next section shows how the researcher applied ISD in a mobile healthcare tool (Bakkar, 2016).

BEST TRAINING PRACTICES IN MOBILE HEALTHCARE

The development of the instructional materials were based on the patient and family rights (PFR) standard as the instructional content employed for the PhD experiments (Bakkar, 2016). Merrill's first principles of instructions selected as the best instructional strategy to implement each instructional-task and to highlight the required instructional-objective. After reviewing existing mobile healthcare (mHealthcare) instructional materials and available off-the-shelf software, the researcher built new PFR instructional content. Microsoft Office PowerPoint employed for the digital-materials' collection and to design the training interface/screens. The Microsoft Office 2007 Clip Art developed most of the graphical-illustrations. The role-play vignettes were videoed by the researcher and inserted to support and explain the instructional-tasks. The these digital-materials (artifacts), such as the videos/vignettes, pictures and textual objects were all imported into the PowerPoint template, which had been structured/designed and built, based on Merrill's first principles of instruction (Bakkar, 2016 pp.130-131).

This design process took into account the fact that the training delivery would occur only via an iPad; where there was no role for any instructor, and the trainees undertook their training at their own pace. Thus, the instructional materials developed combined all the figures, illustrations and videos/vignettes to meet the mHealthcare training objectives, through an accurate and clear logic flow of ideas. The instructional materials were brought to life using the Apple Xcode 5.0 platform for mobile (software) applications development (Bakkar, 2016 p.131).

Table 1. Thematic structure of the mHealth training ID (Bakkar, 2016 p.119)

PFR-1-Task-2 as Merrill's ID principles	Patients are protected from physical assault
RULES: **Just the Facts**	The organisation needs to protect patients from physical assault by visitors, other patients, and/or staff.
VALUES & PRINCIPLES: **Behind the Scene**	**Values:** Protect patients from physical assault. **Principle:** It is wrong not to protect patients from physical assault. **Policy:** Each organisation establishes effective processes and quality procedures to protect patients from physical assault. **Audited activity:** Healthcare staff needs to comply with the law to avoid being sued.
EXAMPLES: **It is like this**	It is the patient's right to be free from such things as mental, physical, sexual and verbal abuse, exploitation, and harassment.
NON-EXAMPLES: **It is never like this**	Hospital staff should never restrain or impose non-medical procedures in any form, or use patients as a means of coercion, discipline, convenience, or retaliation by staff. The consequences of these actions are: • Patients feel uncomfortable, and • Emotional pain and/or healthcare side effects.
KIND of LESSON: **It is kind of like this**	Such protection enables the patient to trust that their life is in safe and secure hands.
PART of LESSON: **It is part of this**	Holistic patient protection from physical assault is part of their natural expectation of safety.
TRY ME: **Have a go**	Question
SHOW ME: **Case study**	Case study

The practical demonstrations used in this research study used consistent instructional design (ID) examples and non-examples for each task. For instance: healthcare procedures were demonstrated using case studies enhanced with pictures and role-play videos/vignettes. The demonstration should be consistent with the instructional outcome regardless of the media or the usage of the trainee/learner guidance (Merrill, 2002; Merrill, 2002c). To this end, Figures 9 and 10 show the demonstration consistency applied using examples and non-examples. Figure 9 shows the same idea using a case study approach (Bakkar, 2016 pp.120).

Learner guidance, such as directing the learner to the relevant information, were adopted demonstrating multiple illustrations; explicitly comparing multiple demonstrations forces the learners to take a broad perspective according to Merrill (2002). Learners were guided how to use the mHealthcare training programme using a screen-based tutorial. Arrow buttons clearly directed the learner to the relevant information for each task; illustrations such as pictures, role-play videos and case studies add further richness to the training demonstrations. Figures-12 and 13 show the use of arrows to direct the trainee/learner's attention to the relevant screen-based

Figure 9. The demonstration consistency applied using examples
Source: Bakkar (2016, p.120)

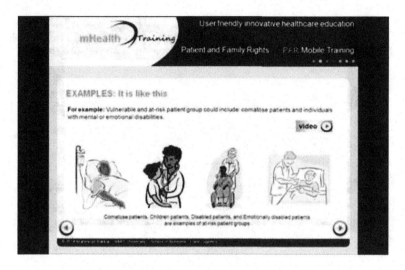

Figure 10. The demonstration consistency applied using non-examples
Source: Bakkar (2016, p.121)

information and how the trainee/learner-guidance was implemented in the tutorial section (Bakkar, 2016 pp.122).

Merrill says the use of varied media in the instructional strategies will promote training/learning and the trainee/learner's attention should not be overwhelmed with varied forms of media (M David Merrill, 2002). Therefore, the iPad training

Figure 11. The demonstration consistency applied using case studies
Source: Bakkar (2016, p.121)

Figure 12. Learner guidance in the tutorial section
Source: Bakkar (2016, p.122)

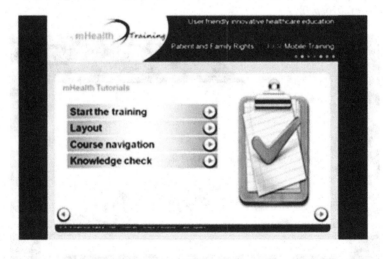

programme consistently used multiple media strategies/resources such as: audio clips; role-play videos/vignettes, and screen-based graphics. Figure-9 shows this use of relevant media in the demonstration phase.

As already noted, the learners' use of new knowledge and skills to solve problems promotes the learning process (M David Merrill, 2002). Moreover, performing real-world tasks involves problem-based model and using practices with the information and the examples will increase the learning effectiveness (M David Merrill, 2002).

Figure 13. The use of arrows to direct the learner to the relevant information
Source: Bakkar (2016, p.123)

The practice should be consistent with the ID ultimate goals and match each type of knowledge and the targeted skill (M David Merrill, 2002) (Bakkar, 2016 p.124).

The stated instructional objective should be consistent with the training-practice and the post-test. The hands-on practice activities should be described with enough information to make it simple for the trainee/learner to recall the practice information, to: locate; describe; name each part of the practice; provide new examples for each kind of practice; know how to do the practice procedures and the consequences of the practice (M David Merrill, 2002). The iPad training programme used the 'kinds-of-lesson: it is kind of like this' section as practice information, linked with a second sequenced section (the 'parts-of-lesson: it is part of this' instructional-item) as part of the thematic structure describing each targeted task. The consequence of following the opposing non-example is shown in the 'non-examples: it is never like this' instructional-item. Figure-14 shows the 'kinds-of-lesson: it is kind of like this' instructional-item in the Merrill's application phase Merrill says the use of varied media in the instructional strategies will promote training/learning and the trainee/learner's attention should not be overwhelmed with varied forms of media (M David Merrill, 2002). Therefore, the iPad training programme consistently used multiple media strategies/resources such as: audio clips; role-play videos/ vignettes, and screen-based graphics. Figure-9 shows this use of relevant media in the demonstration phase (Bakkar, 2016 p.125).

As already noted, the learners' use of new knowledge and skills to solve problems promotes the learning process (M David Merrill, 2002). Moreover, performing real-world tasks involves problem-based model and using practices with the information and the examples will increase the learning effectiveness (M David Merrill, 2002).

The practice should be consistent with the ID ultimate goals and match each type of knowledge and the targeted skill (M David Merrill, 2002).

The stated instructional objective should be consistent with the training-practice and the post-test. The hands-on practice activities should be described with enough information to make it simple for the trainee/learner to recall the practice information, to: locate; describe; name each part of the practice; provide new examples for each kind of practice; know how to do the practice procedures and the consequences of the practice (M David Merrill, 2002). The iPad training programme used the 'kinds-of-lesson: it is kind of like this' section as practice information, linked with a second sequenced section (the 'parts-of-lesson: it is part of this' instructional-item) as part of the thematic structure describing each targeted task. The consequence of following the opposing non-example is shown in the 'non-examples: it is never like this' instructional-item. Figure 14 shows the 'kinds-of-lesson: it is kind of like this' instructional-item in the Merrill's application phase (Bakkar, 2016 p.125).

Figure 15 shows 'parts-of-lesson: it is part of this' section in the application phase.

Figure 17 shows the consequence of demonstrating the opposing non- example section in the Merrill's application phase.

Guiding the learners in their hands-on demonstration of their problem solving promotes the instructional process. This type of guiding can take the form of feedback, with coaching (that is gradually reduced), and consistent error detection and timely correction (M.D. Merrill, 2002a, 2002b). The training application includes a 'try me: have a go' section for each task/theme; posing a question for the trainee/learner to answer. An example of this type of error-detection-screen to depict the correct answer, see Figure 18; while the wrong answer screen is shown in Figure 19.

Involving the learner in solving a sequence of different types of problems promotes the learning process, according to Merrill (2002). The iPad training programme

Figure 14. Kinds-of-lesson section in the application phase
Source: Bakkar (2016, p.125)

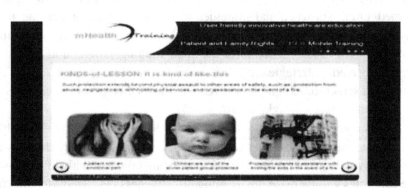

Figure 15. The parts-of-lesson section in the application phase
Source: Bakkar (2016, p.125)

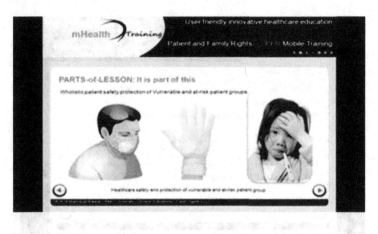

Figure 16. The consequence of demonstrating
Source: Bakkar (2016, p.125)

used the 'show me' instructional-item as part of each task to provide a case study approach including questions for the trainee/learner to answer. Figure-11 shows this 'show me – case study' instructional-item in the iPad's theme structure for each task.

Learners can integrate learning instruction into their daily lives if they have demonstrated an improvement in their skills, retain their new knowledge and adjust it for use in their daily lives (M David Merrill, 2002). Moreover, the learners' newly acquired skills can be refined if they show them to the public or their own close friends (M David Merrill, 2002). The iPad training programme used the 'show me' section to provide a case study and pose questions to the trainee/learners. Figure-11

Figure 17. The 'try me: have a go' item in Merrill's application phase
Source: Bakkar (2016, p.127)

Figure 18. The 'try me: have a go' error detection - correct answer
Source: Bakkar (2016, p.126)

shows an example of a 'show me – case study' instructional-item. The learners were asked questions related to a case they might face in their daily work; in this way, they integrated their learned skills into their own practice and give the case study as an example for their friends or public viewing platform (Bakkar, 2016 p.129).

Figure 19. The 'try me: have a go' error detection - wrong answer
Source: Bakkar (2016, p.126)

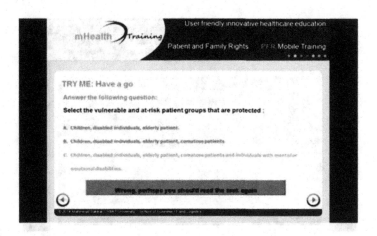

Giving the learners an opportunity to reflect on what they have learned and share it or defend it promotes their acquisition of the new knowledge and skills (M David Merrill, 2002); In this mHealthcare training programme, a feedback input-form enables the learners to suggest necessary changes in the courseware content or instructional-method. Figure 20 shows this feedback button in the homepage screen.

Taking the new knowledge and encouraging modification by the trainees/learners to act on their own instructional-initiatives, makes the trainee/learner transfer their newly acquired knowledge to their own world (M David Merrill, 2002).

The Bakkar (2016) material above showed examples of ISD for delivery of a training programme on a mobile tablet/iPad device. This work commenced with a full task analysis (Dick, Carey, & O'Carey, 2004);(Mager, 1988); and presented a fine example of how to take account of the full range of the human-computer interaction (HCI). The machine-dimension includes all the technical requirements to bring about a successful software application build; while the human-dimension relates to the user's needs (McKay, 2008). Next is the most recent addition to the ISD community as it provides an extension of earlier ISD models to embrace the current multi-media ICT tools.

Figure 20. The feedback button in the homepage screen
Source: Bakkar (2016, p.130)

EVALUATING INFORMATION SYSTEMS
COURSEWARE DESIGN PROCESS

The HE sector continues to embrace emerging ICT tools to reach a wider-customer base aiming for productivity-gains or cost reduction. However, alarming calls exist in the literature of the education technology reveal that high-impact effective eLearning applications are rarely reported (Allen, 2016). Further investigation for the ineffectiveness of eLearning draws the attention to the limitations of existent ID-models as being mainly developed to guide the practice of specific tasks (Young 2008) without referring to educational information systems (IS). It seems possible that these conclusions contribute to the apparent lack of empirical evidence and rigorous ratification processes to measure the effectiveness of these models. A review of the existing accounts support this claim. While there is a plethora of studies that propose 'conceptual' and 'procedural' IS-design, only a few in the field focus on validating these models under different instructional environments (Branch & Kopcha, 2014). Thus, the ultimate goal for this research project was to evaluate the effectiveness of the courseware design process (ISD) to develop instructional materials for an educational IS course for HE and delivered by three different modes of instruction (traditional face-to-face classroom facilitation; blended face-to-face and computerised instruction; fully computerised instruction). The term 'high-impact effective eLearning' defined in this research in terms of the learners' ability to acquire knowledge or improve their cognitive performance. To substantiate the proposed educational IS-design model, the researcher took a systematic approach to investigate how the course delivery mode interacted with a learner's cognitive preference, thereby affecting their performance.

Prescriptive IS-Design Model

A prescriptive IS-design model (McKay, 2018) for this study developed on the Branson ID model incorporating all core ISD design stages (Branson, 1975), which involved: analysis; design; development; implementation and evaluation (Figure-21). The analysis stage describes essential steps and associated techniques required to lay the foundation for the design stage. The design phase articulates three major respects: instructional materials, delivery modes, and learning measures (assessment tests). The focus of this investigation was on the interaction between two elements from both stages (see red-boxes in Figure-21): (1) learners' cognitive preference (Analysis stage); and (2) delivery mode (Design stage) and their effects on performance. While the development of instructional materials and test-items conducted during the third stage, the novelty of this prescriptive educational IS-design model manifested in the systematic validation procedure conducted during the final stage (implementation and evaluation). It aimed to rectify the fidelity of practical aspects initiated with a trial study, and proceeded to three sequenced quasi-experimental studies (pilot studies) prior to the main experiment. The validation process consisted of several activities: plan; execution; observation; preliminary data analysis; refinement; results-recording; and critical reflection. This orderly ISD process enabled the documentation of the necessary practical delineations for effective implementation of the model within a HE context.

Learners' Cognitive Preference

The CSA used in this research identified the cognitive preference of learners (described earlier in this the Introduction of this paper). The CSA test conducted two months prior to the experiments; results used to allocate randomly the participants into one of the instructional treatments: conventional F-to-F classroom facilitation (T1); blended face-to-face and computerised instruction (T2); and wholly computerised instruction (T3). Figure-22 illustrates the CSA results used for the allocation of participants into instructional treatment.

Experimental Procedure

There were a series of 2x3 factorial quasi experiments conducted at four HE institutions during different phases of this research project. The total number of participants was 167 undergraduates who voluntarily participated in this study. Figure 23 shows

Figure 21. Prescriptive educational IS-design model

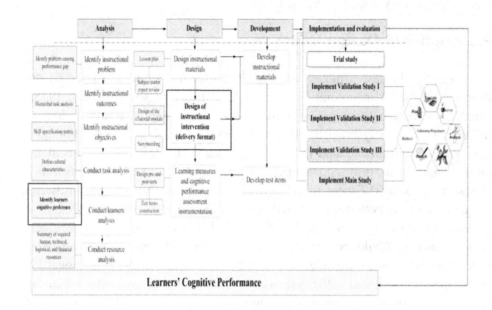

Figure 22. Allocation of participants into instructional treatment based on their CSA results

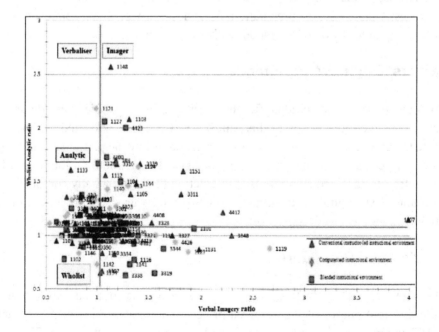

the staged experimental procedure including four key steps conducted during the participants' prescribed tutorial time.

The Design of the eTutorial Module

Accommodating to the needs of learners with different cognitive preferences, some interactivity features were included in the design of the eTutorial module. To this end, two navigation bars were located on the computer-screen in two positions: one located in the navigation bar at the bottom of the screen to allow smooth movement between the different module parts; and a general knowledge navigator button located at the left-side of the computer-screen to enable users to repeat a particular task, or to choose certain other parts of a particular module. Further, the instructional materials presented in the form of: screen-based textual blocks; diagrams and pictures; in a combination of both to suit the verbal and imagery preferenced thinkers (Figure 24). Colours were used to highlight critical parts of the instructional system to provide learners with some support with the structure, should they need this, responding to the preference diversity of the learners.

Preliminary Performance Analysis

Performance measurement and evaluation was conducted using the IRT-based QUEST software tool referred to in the second (McKay & Izard, 2014) project. This useful Rasch measurement tool provided the performance measurement of test-items (questions' relative behaviour) versus participants' relative performance along a unidimensional logit scale (the variable map output – Figure 25) enabled accurate performance inferences (example shown is from Barefah & McKay, 2018).

It also visualized the reliability-fit of test items conforming to the Rasch requirements (via item-fit map depicted in Figure 26 Barefah & McKay, 2018) which was utilised as a validity indicator.

Figure 27 shows further reliability indicators for each test-item taken from the QUEST 'item analysis table.'

QUEST kid-maps show the performance for each individual participant with relevance to the achieved and not achieved test-items. This learning analytics feature was utilised in this investigation to denote the variation of performance for different participants who received instruction with three delivery modes (T1, T2 and T3 described earlier in this section).

To investigate the extent of treatment effects on participants' performance, Cohen's effect size calculated quantifying the size of difference between the experimental groups.

Figure 23. The experimental procedure

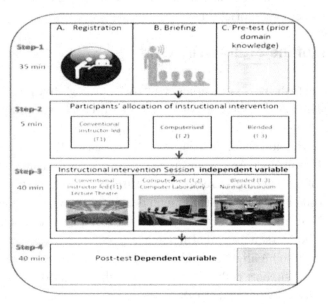

Figure 24. Screenshots of the eTutorial modules

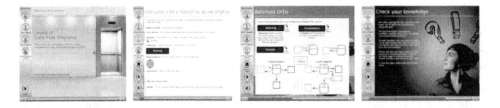

Overall, preliminary results provide reliable evidence regarding the validation of the prescriptive IS-design model under three key delivery modes. The utilization of the Rasch measurement indicators has established the validity and precision of the measuring scale, and allowed measurable performance outcomes. Also, comparable results between the performances of different groups involved in this investigation enabled the delineation of ISD guidelines to improve the ePedagogical practices within HE.

This study is offered as the most recent contribution to the ISD community in that at its core are the ingredients that extend the current view of ISD with an IS-Design lcns. It has truly brought together all the methodological features in the previous projects given earlier in this chapter.

Figure 25. QUEST variable maps (pilot study-3, pre-and-post-tests)
Source: Barefah and McKay (2018)

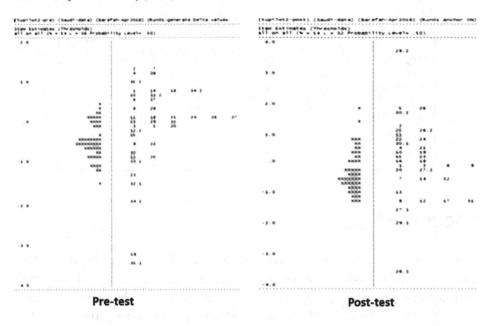

Pre-test   **Post-test**

Figure 26. Item fit maps
Source: Barefah and McKay (2017)

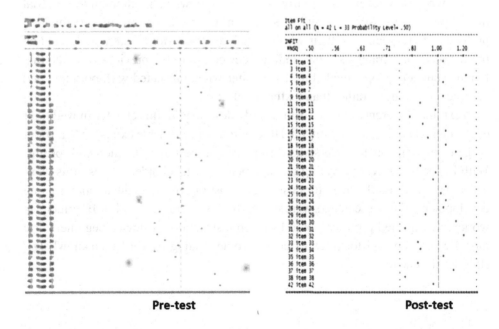

Pre-test   **Post-test**

Figure 27. Extract of QUEST item analysis table for item 30
Source: Barefah and McKay (2018)

```
Item    30: item 30                           Infit MNSQ = 1.14
                                                     Disc =  .33

Categories          0           1           2     missing

Count              28          13           0           0
Percent (%)      68.3        31.7          .0
Pt-Biserial      -.33         .33          NA
p-value          .018        .018          NA
Mean Ability     1.11        2.15          NA          NA

Step Labels                    1

Thresholds                  2.40
Error                        .38
```

DISCUSSION

The main features of this chapter reveal the evidence from each of the researchers standing on the shoulders of others to advance the ID community bringing their work forward into an evolving and dynamic discipline. Mohamad (2014) used both the CSA and the Rasch IRT model to identify the interactive effects of online instruction and cognitive (thinking) preferences. Her findings showed that people who are good at analytic tasks and have a preference to think in terms of textual mode (concentrating on the words for instance) retain information presented to them longer, when they receive a cognitive signal that draws their attention to essential information in the lesson. In addition, novice learner/trainee with an imagery thinking preference (irrespective of their wholist-analytic dimension) were not affected by the instructional strategies. This finding goes contra to the previous research that found (younger) people tended to learn better when presented with both text and graphical metaphors, rather than textual mode alone.

Next, the researchers in the second study described in this chapter moved their experimental context away from the HE environment to enable exposure to a wider range of participants to understand differences in these two instructional forums. With industry partners providing management key stakeholder interest, this work was able to address the need for customised cost of effective educational/training development. These researchers were successful in showing their online ID principles worked as expected providing both lockstep instruction for novice/beginners or a more flexible approach for the experienced people wishing to refresh their knowledge/skill development.

Following on with mobile courseware training development seems a natural progression of ISD. The Bakkar (2016) project again was set in a corporate environment rather than HE to reflect the real world. The thesis excerpt chosen for this chapter was to remind us and accentuate the principles of instruction instead of diving willy-nilly into courseware design. Just because we engage in everything mobile, this does not mean we can abandon first principles entirely.

Consequently, the Barefah doctoral study was included in the chapter as a means to point the way forward to include a global approach to receiving instruction online. To this end, the prescriptive IS-Design model presented here as an extension to the ADDIE model, which is sited by many researchers. Yet ADDIE is a generic analysis, design, development implementation, and evaluation strategy. Barefah's work extends ADDIE to bring ISD into a more flexible IS-design mandate that links the design of instructional interventions (the delivery mode) with a learner's cognitive preferences.

CONCLUSION

There can be no doubt that ICT has brought us a tremendous number of ways to deliver our courseware. Along with this feast of multi-media delights is the relative ease which many people experience these days putting their learning/training programmes online. However the truth is in the telling; or the proof of the pudding is in the eating. Think of how many times you have reached for your mobile/lap top to learn something new, only to find it does not provide flexible knowledge navigation (that's ID training jargon instructional pathway), you become frustrated and eventually give up. The key points from this chapter involve the need for researchers to stop resting on their laurels where they punch out the familiar (outdated) ID models, show more initiative, and bring their ISD into 2018 to reflect the way people want to receive their information these days.

REFERENCES

Adams, R. J., & Khoo, S.-T. (1996). *QUEST:The Interactive Test Analysis System* (Vol. 1). Melbourne: Australian Council for Educational Research.

Allen, M. W. (2016). *Michael Allen's Guide to e-Learning: Building Interactive, Fun, and Effective Learning Programs for Any Company*. John Wiley & Sons. doi:10.1002/9781119176268

Allwood, J., Traum, D., & Jokinen, K. (2000). Cooperation, Dialogue and Ethics. *International Journal of Human-Computer Studies*, *53*(6), 871–914. doi:10.1006/ijhc.2000.0425

Alwi, A., & McKay, E. Investigating an online museum's information system: Instructional design for effective human-computer interaction. In P. I. Demetrios, G. Sampson, & J. M. Spector (Eds.), *D.Ifenthaler, Kinshuk*. Germany: Multiple Perspectives on Problem Solving and Learning in the Digital Age.

Bagley, C. A., & Heltne, M. (2003). *Cooperative Teams*. Paper presented at the Collaboration Conference, Radisson South, Bloomington, MN.

Bakkar, M. N. (2016). *An Investigation of Mobile Healthcare (mHealcare) Training Design for Healthcare Empoyees in Jordan*. RMIT University, School of Business Information Technology and Logistics.

Barefah, A., & McKay, E. (2017). Evaluating the effectiveness of teaching information systems courses: A Rasch measurement approach. *Proceedings of the 9th Annual International Scientific Conference (DLCC)*.

Barefah, A., & McKay, E. (2018). A Rasch-driven validation for measuring the effectiveness of eLearning Applications in Higher Education. *Seventh International Conference on Probabilistic Models for Measurement*.

Blazhenkova, O., & Kozhevnikov, M. (2009). The New Object-Spatial-Verbal Cognitive Style Model: Theory and measurement. *Applied Cognitive Psychology*, *23*(5), 638–663. doi:10.1002/acp.1473

Bond, T. G., & Fox, C. M. (2015). *Applying the Rasch Model: Fundamental measurement in the human services* (3rd ed.). New York: Routledge, Taylor & Francis.

Branch, R. M., & Kopcha, T. J. (2014). *Instructional design models. In Handbook of research on educational communications and technology* (pp. 77–87). Springer.

Branson, R. K. (1975). *Interservice Procedures for Instructional Systems Development. Executive Summary and Model*. DTIC Document.

Clark, R. C., Nguyen, F., & Sweller, J. (2006). *Efficiency in Learning: Evidence-based Guidelines to Manage Cognitive Load*. San Francisco: Wiley & Sons.

Cooper-Smith, L. B., & McKay, E. (2015). *A Comparative Analysis of Tools Used to Measure Engagement in Online Collaborative Learning Activities*. Paper presented at the 2015 IEEE 3rd International Conference on MOOCS, Innovation and Technology in Education (MTE), Amritsar, India.

Dick, W. O., Carey, L., & O'Carey, J. (2004). *The Systematic Design of Instruction* (6th ed.). Allyn & Bacon.

Lee, J. (2007). The effects of visual metaphor and cognitive style for mental modeling in a hypermedia-based environment. *Interacting with Computers, 19*(5-6), 614–629. doi:10.1016/j.intcom.2007.05.005

Mager, R. E. (1988). *The New Six Pack: Making Instruction Work*. Lake.

Mautone, P. D., & Mayer, R. E. (2001). Signaling as a cognitive guide in multimedia learning. *Journal of Educational Psychology, 93*(2), 377–389. doi:10.1037/0022-0663.93.2.377

Mayer, R. E. (2001). *Multimedia Learning*. Cambridge, UK: Cambridge University Press. doi:10.1017/CBO9781139164603

McKay, E. (2008). The Human-Dimensions of Human-Computer Interaction: Balancing the HCI Equation (Vol. 3). Amsterdam: IOS Press.

McKay, E. (2018). Prescriptive Training Courseware: IS-Design Methodology. *AJIS. Australasian Journal of Information Systems, 22*.

McKay, E., Axmann, M., Banjanin, N., & Howat, A. (2007). *Towards Web-Mediated Learning Reinforcement: Rewards for Online Mentoring Through Effective Human-Computer Interaction*. Paper presented at the 6th IASTED International Conference on Web-Based Education, Chamonix, France. Retrieved 15/04/07 http://www.iasted.org/conferences/pastinfo-557.html

McKay, E., & Izard, J. (2013). Seamless Web-Mediated Training Courseware Design Model: Innovating Adaptive Educational-Learning Systems. In A. P. Ayala (Ed.), *Intelligent and Adaptive Educational-Learning Systems: Achievements and Trends* (Vol. 17, pp. 417-442). Springer Berlin Heidelberg.

McKay, E., & Izard, J. (2016). *Planning effective HCI courseware design to enhance online education and training*. Paper presented at the 18th International Conference on Universal Access in Human-Computer Interaction, Toronto, Canada. 10.1007/978-3-319-39399-5_18

McKay, E., & Izard, J. F. (2014). *Online Training Design: Workforce reskilling in Government agencies*. Paper presented at the IC3e 2014 IEEE Conference on eLearning, eManagement and Services, Swinburne University, Melbourne, Australia. 10.1109/IC3e.2014.7081244

McKay, E., & Izard, J. F. (2015). *eLearning programme design? Customised for user-centred participation, Virtual paper presented (and published in Proceedings).* Paper presented at the Global Learn 2015: Global Conference on Learning and Technology, Berlin, Germany.

McKay, E., & Izard, J. F. (2015a). *Evaluate Online Training Effectiveness: Differentiate What They Do and Do Not Know.* Paper presented at the 8th International Conference on ICT, Society and Human Beings 2015 (Multi conference on computer science and information systems - MCCSIS), Las Palmas de Gran Canaria, Spain.

McKay, E., & Izard, J. F. (2015b). *Measurement of cognitive performance in introductory ethics: Workforce reskilling in government agencies.* Paper presented at the 4th IEEE Conference on e-Learning, e-Management and e-Services (IC3e 2015), Melaka, Malaysia. 10.1109/IC3e.2015.7403478

McKay, E., & Vilela, C. (2011). Online Workforce Training in Government Agencies: Towards effective instructional systems design. Australian Journal of Adult Learning, 51(2), 302-328.

Merrill, D. M., Barclay, M., & Van-Schaak, A. (2008). Prescriptive Principles for Instructional Design. In J. M. Spector, M. D. Merrill, M. F. J. van Merriënboer, & M. P. Driscoll (Eds.), *Foundations of Educational Technology: Integrative approaches and interdisciplinary perspectives* (pp. 173–185). Taylor & Frances.

Merrill, D. M., Barclay, M., & Van-Schaak, A. (2016). Prescriptive Principles for Instructional Design. In J. M. Spector (Ed.), *Foundations of Educational Technology: Integrative Approaches and Interdisciplinary Perspectives.* Taylor & Frances.

Merrill, M. (2002). First principles of instruction. *Educational Technology Research and Development, 50*(3), 43–59. doi:10.1007/BF02505024

Merrill, M. D. (2002a). Effective Use of Instructional Technology Requires Educational Reform. *Educational Technology, 42*(4), 13–16.

Merrill, M. D. (2002b). First Principles of Instruction. *ETR&D, 50*(3), 43-59. Available from http://www.indiana.edu/~tedfrick/aect2002/firstprinciplesbymerrill.pdf

Merrill, M. D. (2002c). Pebble-in-the-Pond Model for Instructional Development. *Performance Measurement, 41*(7), 39-44. Available from http://mdavidmerrill.com/Papers/Pebble_in_the_Pond.pdf

Minor, J.I. (2014). *Using Knowledge of Transactional Distance Theory to Strengthen an Online Developmental Reading Course: An action research study* (PhD dissertation). Capella. (3614806)

Mohamad, M. (2012). *The Effects of Web-Mediated Instructional Strategies and Cognitive Preferences in the Acquisition of Introductory Programming Concepts: A Rasch Model Approach.* RMIT University, School of Business Information Technology and Logistics.

Mohamad, M, & McKay, E. (2014). Measuring the Effects of Cognitive Preference to Enhance Online Instruction Through Sound ePedagogy Design. *International Journal of Technology Enhanced Learning, 7*(1), 21-25.

Parker, N. K. (2004). The Quality Dilemma in Online Education. In T. Anderson & F. Elloumi (Eds.), *Theory and Practice of Online Learning* (pp. 385 - 409). Athabasca University. Retrieved 11/05/06 http://cde.athabascau.ca/online_book/copyright.html

Riding, R., & Douglas, G. (1993). The effect of cognitive style and mode of presentation on learning performance. *The British Journal of Educational Psychology, 63*(2), 297–307. doi:10.1111/j.2044-8279.1993.tb01059.x PMID:8353062

Riding, R. J. (2005). *CSA Making Learning Effective - Cognitive Style and Effective Learning.* Learning & Training Technology.

Wu, M., & Adams, R. (2007). *Applying the Rasch Model to Psycho-social Measurement: A practical Approach.* Academic Press.

Young, P. A. (2008). Integrating culture in the design of ICTs. *British Journal of Educational Technology, 39*(1), 6–17.

Yu-Chen, H. (2007). The Effects of Visual Versus Verbal Metaphors on Novice and Expert Learners' Performance. In J. A. Jacko (Ed.), *Human-Computer Interaction Part IV HCII* (pp. 264–269). Berlin: Springer-Verlag.

Chapter 4
Online Social Capital Among Social Networking Sites' Users

Azza Abdel-Azim Ahmed
Abu Dhabi University, UAE & Cairo University, Egypt

ABSTRACT

This research aimed to explore types of online social capital (bridging and bonding) that the Emiratis perceive in the context of social networking site (SNS) usage. A sample of 230 Emiratis from two Emirates, Abu Dhabi and Dubai, was used to investigate the hypothesis. The results showed that WhatsApp was the most frequent SNS used by the respondents. Also, a significant correlation of the intensity of social networking usage and bridging social capital was found, while there was no significant association between SNS usage and bonding social capital. The factors determined the SNSs usage motivations among the respondents were exchange of information, sociability, accessibility, and connections with overseas friends and families. Males were more likely than females to connect with Arab (non-Emiratis) and online bonding social capital. Both genders were the same in their SNSs motivations and online bridging social capital.

INTRODUCTION

Internet and Social Media Connect People

In geographic communities, people typically get to know each other in face-to-face settings, and then maintain contact via communication technologies, such as telephone and email. When geographic communities have high Internet penetration, people, groups, and organizations readily turn to email and the World Wide Web

DOI: 10.4018/978-1-5225-5715-9.ch004

to stay in touch and exchange information (Kavanaugh, et al., 2005). Early and continuing excitement about the Internet saw it as a stimulating positive change in people's lives by creating new forms of online interaction and enhancing offline relationships. The Internet would restore community by providing a meeting space for people with common interests and overcoming the limitations of space and time (Wellman, et al., 2010: 438).

Gershuny (2002) argued that the Internet has changed the nature of leisure activities; the same might be said of social networking sites (SNSs). Instead of displacing leisure or communication, Facebook constitutes a new communication activity that supplements communication amongst friends. Social media provides individuals an interpersonal connection with others, relational satisfaction, and a way to learn about the surrounding cultural milieu (Croucher, 2011: 261). Online sites are often considered innovative and different from traditional media, such as television, film, and radio, because they allow direct interaction with others (Pempek, et al., 2009: 229).

To summarize, SNSs provide users with meaningful ways to make, maintain, and enhance relationships. For many "Friends", the site is the primary method through which to stay connected (Vitak, 2012: 469).

Internet Access and Social Media in the United Arab Emirates (UAE)

Social networking has spread around the world with remarkable speed. In countries such as Britain, the United States, Russia, the Czech Republic, and Spain, about half of all adults now use Facebook and similar websites (Kohut, et al., 2012: 1).

The UAE has been ranked (13) in the world in terms of individuals using the Internet, with 88% of the country's residents now online. This is just behind the United Kingdom (89.8%) and Bahrain (90%), according to the United Nations Broadband Commission report (2014), which elaborates on the number of Internet users, specifically broadband, in 191 countries. In global rankings of countries with the highest frequencies of Internet access, the UAE holds 13th place, way ahead of United States, which is in the 19th spot, and Germany, which has grabbed the 20th position (p: 102–103). It should be indicated here that the demographics of UAE residence are very unique as it includes various nationalities from Asia, Europe, USA and others. According to the National Bureau of Statistics (2010), the UAE nationals are 11.5% of the total population that exceeds 8.264.070 million and the non-national are 88.5% of it (P: 10).

Ayyad (2011) indicated that the United Arab Emirates' high percentage of Internet users makes it the "most wired nation in the Arab world and one of the top nations of the online world" (p: 43).

Al Jenaibi (2011) concluded that social media has a very strong presence in the lives of a sample of 556 Emiratis from the seven Emiratis of UAE. Most participants agreed that the use of social media is on the rise in the current teenage and adult population (Twitter, YouTube, the iPhone, Blackberry, and iPad were mentioned frequently). They had a clear conception of a wide range of uses for it, defining it as useful for contacting others, discussions, searching for information, selling products and logos, making announcements, and distributing surveys (p: 19, 20). Wiest and Eltantawy (2012) found that nearly 90% of a sample of UAE universities' students have created a profile on one of the social networking sites and 78.5% have a profile on more than one such site (p: 214). Karuppasamy, et al. (2013) found that most of a sample of the students of Ajman University of Science and Technology (n = 300) were found to be users of social networking sites, and Facebook was the most popular SNS (p: 248).

THEORETICAL FRAMEWORK AND LITERATURE REVIEW

Like traditional media, Facebook and other social networking sites consist of a one-to-many communication style, where information presented reaches many "viewers" at a time. However, with social networking sites, users are now the creators of content, and they view one another's profiles and information rather than viewing mass-produced content made by large corporations. They also become the stars of their own productions (Pempek, et al, 2009: 234).

Social networking sites are online environments in which people create a self-descriptive profile and then make links to other people they know on the site, creating a network of personal connections. Participants in social networking sites are usually identified by their real names and often include photographs; their network of connections is displayed as an integral piece of their self-presentation (Donath & Boyd, 2004: 72).

Wellman et al (2010) indicated that online interactions may supplement or replace those interactions that previously were formed offline. Some other researchers (Kavanaugh and Patterson, 2001; Hampton and Wellman, 2003; Kavanaugh, et al., 2005) have concluded that computer-mediated interactions have had positive effects on community interaction, involvement, and social capital. Social network sites now mediate a variety of human interactions for a wide spectrum of individuals, from early adolescents to adults (Ahn, 2011: 108). Donath and Boyd (2004) argued that the SNSs provide the technical features for their users to build and maintain large networks for social ties, which supplements their offline social networks. Specifically, individuals can remain in contact with more members of online networks more often than with their offline counterparts.

Social Capital and Its Types

Social capital refers to the set of resources embedded within community networks accessed and used by individuals within a network (Coleman, 1988). Putnam (2000) defined social capital as connections among individuals and the social networks and the norms of reciprocity and trustworthiness that emerge from them. Lin (2001) defined social capital as "investment in social relations with expected returns in the marketplace" (p: 19). Ellison, et al., (2014) explained that social capital is created through social interactions and the expectations of future social resources they engender (p: 856).

Some studies indicated that social capital is linked to positive social outcomes, such as better public health, low crime rates, and increased participation in civic activities (Adler & Kwon, 2002; Ellison, et al., 2007). Wellman, et al., (2010) stated that some evidence suggests that the observed decline in the offline social capital has not led to social isolation but to the community becoming embedded in social networks rather than groups, and a movement of community relationships from easily observed public spaces to less accessible private homes. If people are tucked away in their homes rather than conversing in cafes, then perhaps they are going online: chatting online one-to-one; exchanging e-mail; ranting about important topics; and organizing discussion groups or news groups (p: 437).

Although sociologists and political scientists tend to use the term "social capital", psychologists refer to a related concept using the term "social support" (Burke, et al., 2011, p: 1–2). In media literature, most scholars use "social capital" (Ellison et al. (2007; Ellison, 2008; Valenzuela, et al., 2009; Watkins & Lee, 2009).

Stevens Aubrey, Jennifer et al (2008) summarized Putnam's (2000) distinction of two kinds of social capital: bridging (characterized by weak ties) and bonding (characterized by strong ties).

- Bridging occurs when individuals from different backgrounds make connections between social networks. It is often seen as having a lot of tentative relationships ("weak ties") that provide little emotional support. Still, bridging can also be viewed as the broadening of one's social horizons or world views.
- Bonding, on the other hand, occurs when strongly tied individuals provide emotional support for one another. It occurs between individuals who have strong personal connections. The downside of bonding is its insularity; it can lead to mistrust and dislike for those outside the group (p: 2).

Johnston, et al., (2013) explained that bridging social capital occurs between individuals of different ethnic and occupational backgrounds and it provides useful information and new perspectives (p: 25). Bridging may broaden social horizons or world views, or open up opportunities for information or new resources. On the down side, it provides little in the way of emotional support. (Williams, 2006: 597).

In contrast, bonding social capital exists between family members, close friends, and other close relations and focuses on internal ties between actors. It does not provide links to individuals of differing backgrounds (Johnston, et al., 2013: 25).

Bonding social capital refers to close relationships between individuals that provide emotional support and access to scarce resources. SNSs enable members to connect with existing close friends and relatives, thus functioning as an additional means for them to interact outside of face-to-face encounters (Phua and Annie, 2011: 508). Williams (2006) stated that the individuals with bonding social capital have little diversity in their backgrounds but have stronger personal connections. The continued reciprocity found in bonding social capital provides strong emotional and substantive support and enables mobilization. Its drawback is assumed to be insularity and out-group antagonism (p: 579).

Adding further to this distinction, Johnston et al., (2013) introduced a different classification of social capital retrieved from the work of two researches (Islam et al, 2006; and Fukuyama, 2001). They suggested that social capital can be broken into two classes: cognitive and structural.

- Cognitive social capital is linked to personal aspects, such as beliefs, values, norms, and attitudes. It is also a by-product of cultural norms like religion, tradition, and shared historical experiences.
- Structural social capital is the outwardly visible features of social organizations, such as patterns of social engagement or density of social networks (p: 25).

Ellison et al. (2007) introduced a third type of social capital called "maintained social capital" that is created when individuals maintain connections to their social networks having progressed through life changes. This type of social capital supports the idea suggested by Bargh and McKenna (2004) who stated that the use of technology can assist people to maintain relationships threatened by changes in geographical location. This might be the case when university students use various types of social media to stay in touch with old high school friends and classmates who moved away to join universities in different countries or locations.

In this research, Putnam's (2000) classification of social capital (bridging and bonding) will be adapted.

Social Media and Social Capital

Social networking sites provide an important source of community and thus represent a key source of social capital in the digital age (Watkins & Lee, 2009: 16). In this context, many research efforts were spent to determine whether social media increases or decreases the level of social capital among SNS users.

SNS users were found to enjoy both the development of new relationships and the maintenance of existing relationships online (Walkins and Lee, 2009). Ahmed, Azza (2015) examined offline social support in relation to online self-disclosure. She found that the more respondents have emotional and informational offline support, the less they are likely to disclose positive matters online (p: 215). Nie (2001) argued that Internet use detracts from face-to-face time with others, which might diminish an individual's social capital. From their surveys with undergraduate Facebook users, Ellison, Steinfield, and Lampe (2007) found that Facebook usage interacted with measures of psychological well-being; Facebook intensity predicted increased levels of maintained social capital, which they interpreted as the college students' ability "to stay in touch with high school acquaintances" and possibly "offset feelings of 'friend sickness', the distress caused by the loss of old friends". However, Valenzuela, et al., (2009) found that the positive and significant associations between Facebook variables and social capital were small.

Phua and Jin (2011) suggested that SNSs naturally lend themselves to the development of large heterogeneous networks by enabling individuals to connect with people outside their immediate geographic locations (p: 506). Ahn's (2012) findings suggest that having interactions that are more positive in SNSs is related to bonding social capital but not to bridging relationships. He concluded that when one spends more time in SNSs and interacts with wider networks, one may readily keep in touch with acquaintances rather than developing close relationships, stating that the intensity of SNS use appears to influence bridging social capital development (p: 107).

Ellison, Steinfield and Lampe (2007) found that both strong and weak social ties are sustained on SNSs. Studying a sample of undergraduate university students, they concluded that intensive use of Facebook was associated with higher levels of three types of social capital: bridging capital or our "friends of friends" that afford us diverse perspectives and new information; bonding capital or "the shoulder to cry on" that comes from our close friends and family; and maintained social capital, a concept the researchers developed to describe the ability to "mobilize resources from a previously inhabited network, such as one's high school"(see also: Ellison, 2008: 22).

Greenhow and Robelia (2009) explained that the computer-mediated communication has the potential for online social interactions to enhance self-presentation, relational maintenance, and social bonding (p: 1133). Wong (2012) explained that people might be eager to present themselves in certain ways so as to manage their optimal impressions of others and get social support in return online (p: 185). Koku & Wellman (2001) argued that the Internet may be more useful for maintaining existing ties than for creating new ones. Thus, there is evidence that the Internet plays a critical role in shaping and maintaining bridging and bonding social capital (Phua and Annie, 2011: 506).

Motives of Social Networking Sites Usage and Social Capital

The uses and gratifications approach argues that different audiences use media messages for different purposes to satisfy their psychological and social needs and achieve their goals (McQuail, Blumler, & Brown, 1972). Recently, a number of researchers have employed the uses and gratifications approach in the context of new media and the social networking sites (Dunne, Lawlor & Rowley, 2010). According to Colás, et al., (2013), social network is a virtual space that is emotionally gratifying and allows young people to express their intimate feelings through the perception others have of them (p: 21).

Papacharissi (2002) identified six motives for using social networking sites: "passing time", "entertainment", "information", "self-expression", "professional advancement", and "communication with family and friends". Similar to the results pertaining to traditional media, she concluded that information and entertainment motives were most important. Banczyk, et al. (2008) found that communication and entertainment are the most important motivations for hosting a profile at MySpace, followed by passing time, providing information, and conformity (p: 15). Boyd (2008) found that teenagers were joining SNSs because "that's where their friends are" (p: 126). Karuppasamy, et al. (2013) observed a positive association between meeting new people on SNS and an SNS addiction score; 38.2% of the moderate to high users and 21.2% of the average users used SNS for this purpose (p: 247).

Lenhart and Madden (2007) found that 91% of teen SNS users use the sites "to stay in touch with friends they see frequently and 82% to stay in touch with friends they rarely see in person", whereas only 49% use these sites "to make new friends" (p: 2).

Based on a survey of Facebook users, the findings of Villegas, et al. (2011) suggest that information and connection motives have a positive relationship with the perceived value of advertising on the site. However, when the motivation to use an SNS is moderated by bonding, perceptions toward advertising's value are negative (p: 69).

Aubrey & Rill (2013) found that those who were motivated to use Facebook for its sociability function were more likely to experience gains in online bridging and bonding. They explained that relationship-building on Facebook would be more appealing to the person who is using FB to meet people (sociability) than to the person who is using FB to create an ideal presentation of self (status) (p: 492).

Ellison, Steinfield, and Lampe (2011) found that social information-seeking behaviors, such as using FB to know more about friends and neighbors, are significantly correlated to bonding social capital.

The use of SNSs for relational purposes and the resulting social capital and relationships are culturally driven. American college students held larger but looser networks with a far greater portion of weak ties, whereas their Korean counterparts maintained smaller and denser networks with a roughly even ratio of strong and weak ties. American college students also reported obtaining more bridging social capital from their networks in SNSs than did their Korean counterparts, whereas the level of bonding social capital was not significantly different between the two groups (Choi, et al., 2011).

There is little academic work examining the online social networking and social capital in the Arab world. This study investigates the online social capital (bridging and bonding) among Emiratis using social networking sites. In other words, it investigates the type of social capital developed as a result of using SNSs. It also explores how motivation of using SNSs might make a difference in the type of the perceived social capital among respondents. The differences between males and females in their SNSs usage, type of social capital, and type of motivations are also examined.

RESEARCH QUESTIONS AND HYPOTHESES

Based on the literature, three research questions were formed:

RQ #1: What are the SNSs that are most frequently used by Emiratis?
RQ #2: What are the motives of using online social networking sites among Emiratis?
RQ #3: What is the nationality that Emiratis tend to mostly communicate with via SNSs?

In addition, five hypotheses will be examined based on the literature, as follows:

Research Hypotheses

By using the Internet, people are substituting poorer quality relationships for better relationships, substituting weak ties for strong ties (Kraut, et al, 1998: 1208). Boase, et al. (2006) in a Pew Internet survey showed that online users are more likely to have a larger network comprised of close ties compared to non-Internet users. Ellison, N. et al (2006) found that there was a strong connection between Facebook intensity and high school social capital (p: 25). Therefore, it can be hypothesized that:

H1: Intensity of SNS use is positively associated with individuals' perceived online bridging and bonding social capital.

Aubrey & Rill (2013) indicated that one of the structural properties of users' FB experiences that might predict the relationship between SNS usage and the type of online social capital is the number of relationships formed online (p: 493). Valenzuela, et al. (2009) suggested that individuals with a large and diverse network of contacts are thought to have more social capital than individuals with small, less diverse networks (p: 875). Hofer and Aubert (2013) found a negative curvilinear effect of the number of Twitter followers on bridging and the number of Twitter followees on bonding online social capital. Therefore, it can be predicted that the size of the SNS might predict the type of social capital as follow:

H2: There is a correlation between online social networking size and social capital type (bridging and bonding).

The results of Wellman, et al. (2010) suggest that the Internet is particularly useful for keeping contact among friends who are socially and geographically dispersed, concluding that communication is lower with distant than nearby friends. Also, investigating the relationship between Facebook usage and online and offline social capital, Aubrey & Rill (2013) found that Facebook habit was related to online bridging and offline network capital. They concluded that it is likely how a person uses FB, rather than how much he or she uses it, which is related to online bridging and offline network capital. This means that studying the impact of social networking on social capital should be in the light of the types of usage and patterns of SNS use. One of these patterns might be the diversity of social categories the SNS users have. Thus, it can be hypothesized:

H3: There is a correlation between diversity of social categories (various nationalities and relationships) and social capital type (bridging and bonding).

Literature on social media concluded that college students are motivated to use social media to keep in touch with old friends, sharing artifacts, learning about social events, and gaining recognition and self-expression (Ellison, et al, 2007; Papacharissi, 2002; Raacke & Bonds-Raacke, 2008). Greenhow, Christine and Robelia (2009) found that high school students from low-income families used their online social network to fulfill essential social learning functions. Ellison, et al. (2006) found that using FB to connect with offline contacts and for fun were positively associated with bridging social capital, while using it to meet new people had a negative association among a sample of high school students. These motives did not explain bonding social capital well (p: 25). Burke & Lento (2010) found a correlation between social capital and active contributions to Facebook as compared to passive consumption of other's information. Ellison, et al. (2011) found that using Facebook for information-seeking purposes were positively associated with online social capital. Therefore, it can be hypothesized that:

H4: There is a significant correlation between the motivation of using SNSs and type of social capital.

The differences between males and females have been studied by some researchers. Ayyad (2011) found that male students are more interested in using the Internet to engage in dialogue and chat with their friends and relatives, while female students are more interested in using the Internet to get information that serves their studies and to communicate with their instructors (p: 57). In the light of social capital, Colás, et al. (2013) found that online social networks are a source of resources for young people that are used to fulfill needs, both psychological and social. However, the differences between genders in these variables demonstrate that they play a compensatory role; males generally use them to cover emotional aspects ("to feel good when I am sad") and reinforce their self-esteem ("to know what my friends would say about my photos I upload"), while for young women, the relational function prevails "to make new friends" (p: 20). Therefore, it can be hypothesized that:

H5: There is a significant difference between males and females in their:

1. Motivations of using social networking sites, and
2. Online social capital.

Figure 1 illustrates the hypothesized relationships among the research variables.

METHODOLOGY

Sampling and Data Collection

The sample is composed of only Emiratis. Two out of seven Emirates were selected to draw the sample of the study: Dubai, the leading and most modern Emirate in UAE, and Abu Dhabi, the UAE capital.

A constructed self-administrated questionnaire was used to collect the data. It included 12 questions with various kinds of measurements for the research variables which will be described later. The questionnaire was written in Arabic, as it is the native language of the respondents and it makes it easy for them to fill out the questionnaire. The questionnaire reliability was good ($\alpha = 0.851$).

The data was collected during February-March 2014. Five national students of the Mass Communication Program at Abu Dhabi University assisted in collecting data using Snowball sampling from the two Emirates where they live. The students gained extra credit in the Media Research Methods and Communication Theories courses for this extracurricular activity.

Total of (300) questionnaires were distributed and filled in by respondents. A number (70) of questionnaires were excluded due to various reasons, specifically: uncompleted sheets and inaccurate responses. Therefore, the sample was composed of (230) respondents. They were distributed equally between the two Emiratis; their characteristics are shown in Table 1.

Figure 1. The research variables and the related hypotheses

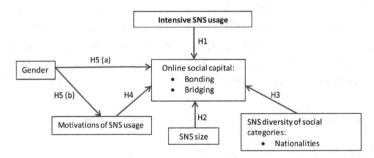

Table 1. The demographics of the sample (n = 230)

Demographic	%
Gender	
• Males	48.3
• Females	51.7
Age (*)	
• Less than 25 years old	49.6%
• 25 to less than 35 years old	38.7%
• 35 years old and above	11.7%
Education	
• University level	67.4%,
• High school level	17.8%
• Post graduate level	10.9%
• Preparatory or less	3.9%

(*) The mean age is 26.13 & St. Deviation 8.143

MEASUREMENTS OF VARIABLES

Types of Social Networks

The respondents were asked about the social networking sites they usually use. A list of social networking sites was provided and the respondents were asked to indicate how frequently (3 "always", 2 "sometimes", 1 "rarely", and 0 "never") they use each of them (Facebook, Instagram, Kik, WhatsApp, BBM, and Twitter).

Intensity of Social Networking Connection

It refers to the frequency of using social media and the time spent on social media. Three questions were used to measure the intensity of the social networking connection adapted from Lee (2009).

The respondents were asked how many years they had been using the social networking sites. The categories were: less than a year, from 1 year to less than 3 years, from 3 years to less than 6 years, and 6 years and above. Due to the small number of respondents who chose "less than one year, this category was combined with the second one, which is "1 year to less than 3 years". The scores ranged from 1–3. The highest score (3) was for the "6 years and above" and the lowest was (1) for the "1 year to less than 3 years" category.

The second question asked about how often they use social media in the average week. The answers were almost every day, 5–6 days a week, 3–4 days a week, once or twice a week, or never. The "3–4 days a week" was combined with the "once or twice a week" due to the low number of respondents who gave this answer. The highest score (3) was for the "almost every day" and the lowest (1) was for "once or twice a week". None of the respondents expressed that they never use the SNS during an average week.

A third question asked about the number of hours the respondents use social networking sites in one day. Possible answers were: less than an hour, 1 to less than 3 hours, 3 to less than 5 hours, 5 hours and more. The first two categories were combined together; so it turned to be "1 to less than 3 hours". The highest score (3) was for the "5 or more" category, followed by 3 to less than 5 hours a day (2) and the lowest (1) was for less than 1 to 3 hours a day.

The total score of this variable was 9 and ranged from 3–9 points. Responses were divided into three categories: highly connected with SNS (8–9 points) 33.9%, moderately connected (5–7 points) 60.4%, and weakly connected (3–4 points) 5.7%. Cronbach's Alpha indicated a good internal reliability of 0.823.

Online Network Size

It refers to the size of friends the respondents have via social networking sites. The respondents were asked how many friends they have in the online social networking sites. The answers have five categories: 50 to less than 100 friends, 100 to less than 200 friends, 200 to less than 300 friends, and 300 friends and more. The results revealed that 46.5% of the sample has more than 300 friends in the SNS while 53.5% has 50 to less than 100 friends.

Diversity of Social Categories

It refers to the diversity of social categories and type of friends in terms of relationship and nationalities. It also reflects the geographical and psychological distance of SNS friends. Two questions were asked to measure this variable. The first asked how frequently the respondents communicate with the following social categories (family, colleagues, friends, relatives, work partners, and strangers). The score ranged from 3 "always", 2 "sometimes, 1 "rarely", and 0 "never". The total score for this part was 18, and respondents were divided according to their responses into three categories: high diversity of relationships (scores from 14–18) 37%, medium diversity (7–13) 60.4%, and low diversity (0–6) 2.6%.

The second question was about the nationality of the friends and followers with which the respondents communicate (Emiratis, non-Emirati Arabs, Americans and Europeans, and Asians). The total score for this question was 15, ranged from 3 "always", 2 "sometimes, 1 "rarely", and 0 "never".

The responses were divided accordingly into three categories: high diversity of nationalities (scores from 12–15) 5.2%, medium diversity (6–11) 58.7%, and low diversity (0–5) 36.1%. Cronbach's Alpha indicated good internal reliability (α = 0.851) for the items measuring the diversity variable.

Motivation

Motives of using the online social networking sites were measured by using a 5-point Likert scale, ranging from strongly agree (4 points) to strongly disagree (0 point) for 13 different motives: "to pass time" (M = 4.03); "to connect with new friends" (M = 3.83); "to find new friends" (M = 3.13); "to have accompaniment" (M = 3.33); "to connect with my friends overseas" (M = 3.97); "to connect with my family living abroad" (M = 3.97); "to exchange opinions with friends" (M = 4.01); "to write comments on everyday events" (M = 3.62); "to upload some videos and photos" (M = 3.74); "to post some up-to-date news that might grab my friends' attention" (M = 3.82); "to read what others have posted" (M = 4.12); "because it is the best way to find the latest news" (M = 4.23); and "because it is the easiest way to connect with my friends" (M = 4.14).

Factor analysis was used to identify the factors that determine the respondents' motivations of using social media (see the results of the research).

Social Capital Measurements

The measurement of the social capital variable was adopted from Chua, Shu-Chuan and Choib, Sejung Marina (2010). Minor changes have taken place while translating the statements as a result of the pre-test and a necessity of changes to go along with the cultural differences between the western and the Arab communities.

The social capital variable consists of two sub-variables: bridging social capital and bonding social capital. The 5-point Likert Scale ranged from "strongly agree" (5) to "strongly disagree" (1 score) was used to measure the social capital variable using (10) statements for each variable as follows:

Bridging Social Capital

1. Interacting with people on the social network site makes me interested in things that happen outside of my town (M = 3.98).

2. Interacting with people on the social network site makes me want to try new things (M = 4.02).
3. Interacting with people on the social network site makes me interested in what people unlike me are thinking (M = 4.04).
4. Talking with people on the social network site makes me curious about other places in the world (M = 3.80).
5. Interacting with people on the social network site makes me feel like part of a larger community (M = 4.06).
6. Interacting with people on the social network site makes me feel connected to the bigger picture of my community (M = 4.08).
7. Interacting with people on the social network site reminds me that everyone in the world is connected (M = 4.03).
8. I am willing to spend time to support general community activities on the social network site (M = 3.82).
9. Interacting with people on the social network site provides me with a chance to connect with new people to talk to (M = 3.77).
10. I come in contact with new people on the social network site all the time (M = 3.69).

The total score was 50 and the respondents were divided according to their answers into three categories: High in bridging social capital 59.1% (scores from 39–50), moderate in the bridging social capital 39.6% (scores from 26–38) and low in bridging social capital 1.3% (scores from 10–25). The low and the moderate categories were combined because of the low percentage of the low category. The respondents then divided into two categories: high in bridging social capital (59.1%) and middle in bridging social capital (40.9%).

Bonding Social Capital

1. There are several members of the social network site I trust to help solve my problems (M = 3.35).
2. There is a member of the social network site I can turn to for advice about making very important decisions (M = 3.33).
3. There is no one on the social network site that I feel comfortable talking to about intimate personal problems (R) (M = 3.27).
4. When I feel lonely, there are members of the social network site I can talk to (M = 3.39).
5. If I needed an emergency loan, I know someone at the social network site I can turn to (M = 2.79).

6. I always evaluate the people I interact with on the social network site (M = 3.47).
7. The people I interact with on the social network site would be good job references for me (M = 3.00).
8. The people I interact with on the social network site would share all their money to support me if I face any financial problem (M = 2.70).
9. I do not know members of the social network site well enough to get them to do anything important (M = 3.57). (R)
10. The people I interact with on the social network site would help me fight an injustice or unfairness that I might face in my life (M = 3.05).

The total score of bonding social capital was 50. The respondents were categorized according to their responses into three categories: high in bonding social capital 16.5% (scores 39–50), moderate in bonding social capital 64.3% (scores 26–38), and low in bonding social capital 19.1% (scores 10–25).

Cronbach's Alpha indicated a high internal reliability (α = 0.851) for the items measuring the social capital (bridging and bonding) variable. The respondents' age, gender, and education levels were recorded for sampling demographics.

Statistical Techniques

The SPSS statistical program was used in analyzing the data. Frequencies, Cronbach's Alpha, Pearson correlation, Factor Analysis, and T-test were used to answer the research questions and test its hypotheses.

RESEARCH RESULTS

Usage Patterns of Social Networking Sites Among the Respondents

RQ #1: What is the SNS/s that are most frequently used by Emiratis?

The results revealed that WhatsApp is the social network that got the highest percentage among SNS Emiratis users as shown in Table 2.

Table 2 shows that the SNS that is used most frequently by the sample is WhatsApp (M = 2.70), followed by the Instagram (M = 2.53), then Twitter (M = 2.08). Unexpectedly, Facebook was in fourth place among the most frequent social networking sites the respondents use. This contradicts some findings of research results in UAE and USA, as will be discussed later.

Table 2. Frequency of using SNS among respondents (n = 230)*

SNS	Frequency of Using SNS %				M	Weighted Mean %
	Always	Sometimes	Rarely	Never		
WhatsApp	81.3	12.2	2.2	4.3	2.70	90.1
Instagram	72.2	15.7	4.8	7.4	2.53	84.2
Twitter	50.9	21.7	11.7	15.7	2.08	69.3
Facebook	35.7	17.0	10.0	37.4	1.51	50.3
Kik	24.3	17.4	22.6	35.7	1.30	43.5
MySpace	11.7	6.5	11.7	70.0	0.60	20.0

(*) The respondent might choose more than one SNS

RQ #2: What are the motives of using social networking sites among Emiratis?

Factor analysis was used to identify the most factors that determine the social media usage among the respondents.

The results indicated that there are four factors identifying the motives of using social media. Table 3 presents the factor analysis rotated component matrix of SNS motivations among Emiratis.

The factor analysis reveals that there are four factors that determine the Emiratis' motivations of using SNS.

1. Exchange news, opinion and photos (items 1–5)
2. Sociability (items 6–9)
3. Accessibility (items 10–11)
4. Connecting overseas friends and families (items 12–13)

RQ #3: What is/are the nationality/ies that Emiratis tend to mostly communicate with via SNS?

Connecting to people is one of the main factors that motivate the intensive usage of SNS among various groups.

The respondents were asked with which of the nationalities they communicate most of the time through SNS. The answers are listed in Table 4.

Table 4 indicates that the Emiratis use SNS to always communicate to nationals, as Emiratis got the highest percentage compared to other nationalities, followed by the Arabs (non-Emiratis). This indicates that the SNSs are not used to extend the respondents' networking outside the Arab region.

Table 3. Factor analysis rotated component matrix of SNS motivations among Emiratis

Factors / Motives	Factor 1 Exchange news, opinion, and photos	Factor 2 Sociability	Factor 3 Accessibility (new and easy way of communication)	Factor 4 Connecting to overseas friends and families
1. To exchange opinions with friends	**0.681**	-.044	0.193	0.328
2. To write comments on everyday events	**0.707**	0.239	0.109	--
3. To upload some videos and photos	**0.641**	0.292	0.102	0.160
4. To post some up-to-date news that might grab my friends' attention	**0.804**	0.123	0.217	--
5. To read what others have posted	**0.534**	0.130	0.501	--
6. To pass time	0.046	**0.614**	--	--
7. To connect with new friends	--	**0.715**	--	0.295
8. To find new friends	0.294	**0.743**	--	--
9. To have accompany	0.388	**0.636**	--	0.144
10. Because it is the best way to find latest news	0.222	0.085	**0.789**	0.065
11. Because it is the easiest way to connect with my friends	0.149	0.068	**0.820**	0.063
12. To connect with my friends overseas	0.166	0.147	0.094	**0.809**
13. To connect with my family living abroad	0.043	0.122	--	**0.836**

Table 4. Frequency of communicating with various nationalities via SNS

The Nationality	Frequency of Connecting to Various Nationalities %				Mean	Weighted Mean %
	Always	Sometimes	Rarely	Never		
Emiratis (Nationals)	81.7	15.2	2.2	0.9	2.78	92.6
Arabs (Non-Emiratis)	38.7	40.0	16.1	5.2	2.12	70.7%
Americans	10.0	16.5	22.2	51.3	0.85	28.4%
Europeans	10.0	12.6	22.2	55.2	0.77	25.8%
Asians	--	0.9	--	--	--	--

HYPOTHESES TEST

This research examines five hypotheses. The results are as follow:

H1: Intensity of SNS use is positively associated with individuals' perceived online bridging and bonding social capital.

Pearson correlation was used to test this hypothesis and the result is shown in Table 5.

The result showed a significant correlation between intensity of SNSs usage and bridging social capital r = .200. However, the correlation is not significant between intensive usages of SNSs and bonding social capital.

This indicates that social networking websites are effective in predicting online bridging social capital, in that the SNSs allow Emiratis to establish tentative relationships with new friends. In addition, the intensive usage of SNSs does not predict bonding social capital, which means that Emiratis tend to have weak ties via SNSs rather than strong personal connections.

H2: There is a correlation between online social networking size and the social capital types (bridging and bonding).

The result of Pearson correlation is shown in Table 6.

Pearson correlation revealed a significant weak correlation between online SNS size and bridging social capital r = .155, while there is no significant correlation between online social networking size and bonding social capital.

Table 5. Correlation between intensive usages of SNSs and online social capital

Variables	Online Social Capital	
	Bridging	Bonding
Intensive Usages of SN Connection	.200(*)	0.100 (NS)

* $P \leq 0.001$ (2-tailed) N = 230 *(NS) non-significant*

Table 6. Correlation between online social networking size and the social capital type

Variables	Online Social Capital	
	Bridging	Bonding
Online social networking size	.155(*)	-0.018 (NS)

Note: * $P \leq = 0.02$ NS: Non significant (2-tailed) N = 230

This means that the bigger the size of SNS, the more the bridging social capital and the lesser the online bonding will be perceived by the respondents.

H3: There is a correlation between diversity of social categories (nationalities and relationships) and social capital type (bridging and bonding)

The results in Table 7 showed a positive significant correlation between diversity of "relationships" connection via SNSs and both of bonding and bridging social capital. However, the diversity of "nationalities" was not significantly correlated to bonding social capital while it was significantly correlated to bridging.

H4: There is a significant correlation between motivation of using SNSs and type of social capital.

Pearson correlation revealed that there are significant correlations ($p = .000$) between respondents' motivations of using SNSs and both types of social capital (Sociability r = 0.242), (Connecting overseas friends and families r = 0.242), (Exchange news, opinions, and photos r = 0.345) and (Accessibility "new and easy way of communication", r = 0.322). The correlation between each factor of motivations and the types of social capitals were significant as shown in Table 8.

Table 7. Correlation between diversity of social categories and social capital types

Variables		Online Social Capital	
		Bridging	Bonding
Diversity of social categories in SNS	Relationships	.173(**)	.139(*)
	Nationalities	.141(*)	0.028 (NS)

Note: * P≤0.03 ** P≤ 0.001 (2-tailed) N = 230 *(NS) non-significant*

Table 8. Correlation between motivations and types of social capitals

Variables		Online Social Capital	
		Bridging	Bonding
SNSs Motivation factors	Exchange news, opinions, and photos	0.374(*)	0.210 (***)
	Accessibility (new and easy way of communication)	0.316(*)	0.174 (*)
	Connecting overseas friends and families	0.225(*)	0.141 (**)
	Sociability	0.241(*)	0.196 (**)

Note: * P≤0.000 ** P≤ 0.003 *** P≤ 0.001 (2-tailed) N = 230 *NS) non-significant*

The results showed a significant correlation between all SNS motivation factors and both types of social capital: bonding and bridging. However, the correlation was stronger between motivation and bridging than bonding.

The correlation was stronger with bridging especially for both motivation factors: "exchange opinions and news" and "accessibility". The "exchange news, opinions, and photos" motive strongly predicts both bridging and bonding.

This indicates that using SNSs more for "exchanging news", "connecting families and friends", "sociability", or "accessibility" is associated with increased online bridging and bonding.

H5: There is a significant difference between males and females in their:
1. Motivations of using social networking sites, and
2. Online social capital.

The T-test was used to examine this hypothesis. It showed no significant difference between males and females in their motivations of SNS usage. The results are shown in Table 9.

The T-test shows that there is no significant difference between males and females in bridging social capital in the social networking sites. The difference was significant in bonding social capital (t = 3.840, p = 0.000). The results show that online bonding social capital is higher among males than females.

In addition, the T-test was used to examine if there is any significant difference between males and females in their communication with various nationalities. The results revealed that there is no significant difference between males and females in their connections to Emiratis, Americans, and Europeans. However, there was a significant difference (t = 2.87, *df = 228, p* = 0.004) between them in their connection to Arabs (non-Emiratis). Males (M = 2.29) have more connection with Arabs (non-Emiratis) than females (M = 1.97).

Table 9. Differences between males and females in their social capital

Online Social Capital Type	Gender	Mean	Std. Deviation	t	*df*	Sig. (2-tailed)
Bridging	Males	2.56	0.499	-0.973	228	0.331 (NS)
	Females	2.62	0.487			
Bonding	Males	2.13	0.558	3.840		0.000
	Females	1.83	0.601			

DISCUSSION AND CONCLUSION

This research investigated the association between intensive usage of the social networking sites and online social capital. Also, the correlation between size of social networking friends/followers and the type of online social capital was investigated. The effect of diversity of relationships and nationalities on the type of online social capital was examined. The type of and effect of SNS usage motivation were tested. In this section, a discussion of the research results is presented.

In this research, it was revealed that the WhatsApp is the most frequently used SNS among the respondents. Facebook was the fourth-used SNS. This result is different from some Arab and Western research findings.

According to the Arab World Online report (2013), Facebook is the most popular social network, followed by Google+, and then Twitter. Most respondents in that report stated they have never used the other social networks listed in the AWO survey. The report also indicated 54% of respondents used Facebook more than once a day, while 30% used Google+ at the same frequency. Only 14% of respondents used Twitter more than once a day (13).

Ellison, N. et al (2007) found that university students intensively use Facebook in their everyday lives. Ahmed, Azza (2010) indicated that Facebook is the most popular online social network (65.2%) among a sample of (325) respondents from Egypt and UAE (p: 82). Mourtada and Salem (2011) found that Facebook is the most popular social media technology in the Arab world, with Twitter rapidly gaining in popularity. Al-Jenaibi, B. (2011) found that the most popular social media technologies are, for the most part, the same as in the West: Facebook, video-sharing sites like YouTube, and micro-blogging sites like Twitter. Wiest, J. and Eltantawy, N. (2012) found that Facebook is overwhelmingly the favorite social networking sites among a sample (n = 179) of UAE university students. In her quantitative and qualitative study, Ahmed, Azza (2015) found the mobile application "WhatsApp" was the most common application used among a sample of 313 Arab residents of the United Arab Emirates. Facebook and Instagram followed it. In addition, the qualitative analysis showed that Facebook is the predominant SNS among the majority of the interviewees along with WhatsApp (p: 205).

It seems that the SNS usage habits have changed rapidly throughout the past few years. The smartphone might have an impact on the frequency of using each type among many other factors, such as the culture factors and the features in each SNS.

The literature suggests that the type of relationships within the social network can predict different kinds of social capital.

The results showed that Emiratis tend to communicate with nationals and Arabs (non-Emiratis) more than people from other nationalities. This result supports the findings of Wellman et al (2010) that suggest the Internet is particularly useful for keeping contact among friends who are socially and geographically dispersed. Yet distance still matters: communication is lower with distant than nearby friends (p: 450). Also, Watkins & Lee (2009) concluded that SNS users tend to interact mostly with friends they already know well rather than be friend to complete strangers (p: 22).

Also, the research revealed that the more diverse the social categories of "relationships" and "nationalities", the more bridging social capital can be predicted. Bonding social capital is associated positively with "relationships", but negatively with "nationalities". This means that Emiratis are keen to use SNS to broaden their social networks rather than get/provide emotional support.

The results showed an association between intensity of SNS connection and bridging social capital, while no significant correlation was found between intensity of SNS connection and the bonding social capital. This result supports the Ellison, Steinfield, & Lampe (2007) findings, indicating that bridging social capital was the most valued use of Facebook. They suggested that networking through these sites may help to crystallize relationships that "might otherwise remain ephemeral" (p. 25). They explained that the intensity of Facebook usage can help students accumulate and maintain bridging social capital. This form of social capital; which is closely linked to the notion of "weak ties", seems well suited to social software applications, because it enables users to maintain such ties cheaply and easily.

Granovetter (1973) explained this process by saying that the social capital is to encourage users to strengthen latent ties and maintain connections with former friends, thus allowing people to stay connected as they move from one offline community to another.

Johnston, et al. (2013) results indicate that intensity of Facebook use plays a role in the creation of social capital, but it is particularly significant regarding the maintenance of social capital in the South African context. The Facebook intensity is positively correlated with all types of social capital.

However, Aubrey & Rill (2013) found that FB use was not associated with online bridging or online bonding; they suggested that Facebook might not be effective in facilitating bonding relationships, but it might help users to maintain already established weak ties between individuals (p: 491). Wellman, et al. (2010) concluded that the greater use of the Internet may lead to larger social networks with more weak ties and distasteful interactions with some of these ties, resulting in lower commitment to the online community (p: 494). Phua, and Jin, (2011) found that intensity of SNS usage was significantly positively associated with bonding and bridging social capital in the US college environment (p: 512). In addition, Ahn (2012) found that time spent on Facebook and MySpace was significantly related to bridging social capital, whereas there was no relationship to bonding social capital (p: 106).

In the current study, four factors were found to determine the respondents' motivations of SNS usage. These factors are: Exchange information, Sociability, Accessibility (which refers to easy way to access new information), and Connecting with overseas friends and families. This finding supports the results of Ellison (2007) who reported that Michigan undergraduate students overwhelmingly used Facebook to keep in touch with old friends and to maintain or intensify relationships characterized by some form of offline connection, such as dormitory proximity or a shared class.

In addition, the findings of the current research revealed that the four SNS motivation factors were associated with increase online bridging and bonding. This finding supports Aubrey & Rill (2013) results that found using Facebook for sociability reasons was associated with increased online bridging and bonding. Moreover, they provided evidence of the sociability motivation mediating the relationship between Facebook habits and online bonding and bridging. Ellison, et al. (2011) found that FB communication practices focused on using the site for social information-seeking purposes were positively associated with online social capital. In this context, Donath and Boyd (2004) suggested that Facebook allows users to maintain weak ties cheaply and easily. These findings suggest that intensive users of SNSs are likely to experience gains in online social capital. The results of Haythornthwaite (2000) suggest that online communicators who communicate with each other more frequently may be more likely to have closer relationships; therefore, more frequent communication with other Facebook users is necessary to build closer relationships as a first step towards bonding social capital.

LIMITATION AND SUGGESTIONS

Although the method used in this study does not allow for making a causal conclusion, it was clear that intensive usage of social networking sites is a significant predictor of online bridging social capital among Emiratis. However, the research findings are still not generalizable to the larger Emiratis population of SNS users in other Emirates.

Future studies should investigate how these variables interact among other populations of users. More research efforts should be directed to study how new generations are using the social networking websites and the mechanism used to build their social capital. More research should be conducted to investigate the differences between social capital in online and social offline settings for the two types of social capital in the Arab region. In other words, does online social capital strengthen or weaken the offline social capital? This field of research should attract more Arab scholars to investigate the impact of social media on Arab societies and whether these impacts differ from their counterparts in the US and Europe.

A longitudinal research should be conducted to investigate the habits of SNS usage among Arab societies. The findings of such research might help in planning how social media will be used to affect the Arab youth's attitude and behaviors.

In conclusion, as social network sites grow, understanding the interaction between site features and individual differences in users will become even more important and will require more attention from the media scholars.

REFERENCES

Ahmed, A. A. A. (2010). Online privacy concerns among social networks' users. *Cross-Cultural Communication, 6*, 74–89.

Ahmed, A. A. A. (2015). "Sharing is Caring": Online Self-Disclosure, Offline Social Support, and SNS in the UAE. *Contemporary Review of Middle East, 2*(3), 192–219. doi:10.1177/2347798915601574

Ahn, J. (2012). Teenagers' Experiences with Social Network Sites: Relationships to Bridging and Bonding Social Capital. *The Information Society, 28*(2), 99–109. doi:10.1080/01972243.2011.649394

Al Jenaibi, B. N. A. (2011). Use of Social Media in the United Arab Emirates: An Initial Study. *Global Media Journal Arabian Edition, 1*(2), 3-27. Retrieved from: http://www.gmjme.com/gmj_custom_files/volume1_issue2/articles_in_english/volume1-issue2-article-3-27.pdf

Aubrey, F. S., & Rill, L. (2013). Investigating Relations between Facebook Use and Social capital Among College Undergraduates. *Communication Quarterly, 61*(4), 479–496. doi:10.1080/01463373.2013.801869

Aubrey, S., Chattopadhyay, S., & Rill, L. (2008). *Are Facebook Friends Like Face-to-Face Friends: Investigating Relations Between the Use of Social Networking Websites and Social capital.* Paper presented at the annual meeting of the International Communication Association.

Ayyad, K. (2011). Internet Usage vs. Traditional Media Usage among University Students in the United Arab Emirates. *Journal of Arab & Muslim Media Research, 4*(1), 41–61. doi:10.1386/jammr.4.1.41_1

Banczyk, B., & Senokozlieva, M. (2008). *The "Wurst" Meets "Fatless" in MySpace: The Relationship Between Self-Esteem, Personality, and Self-Presentation in an Online Community.* International Communication Association, Annual Meeting.

Bargh, J., & McKenna, K. (2004). The Internet and Social Life. *Annual Review of Psychology, 55*(1), 573–590. doi:10.1146/annurev.psych.55.090902.141922 PMID:14744227

Boase, J., Horrigan, J., Wellman, B., & Rainie, L. (2006). The Strength of Internet Ties: The internet and E-mail Aid Users in Maintaining Their Social Networks and Provide Pathways to Help When People Face Big Decisions. *Pew Internet and American Life Project.* Retrieved from: http://www.pewinternet.org/files/old-media/Files/Reports/2006/PIP_Internet_ties.pdf.pdf

boyd, d. (2008). *Taken Out of Context: American Teen Sociality in Networked Publics* (PhD Dissertation). University of California-Berkeley, School of Information.

Burke, M., Kraut, R., & Marlow, C. (2011). *Social Capital on Facebook: Differentiating Uses and Users.* CHI 2011, Vancouver, Canada.

Burke, M., Marlow, C., & Lento, T. (2010). Social Network Activity and Social Well-Being. In *Proceeding of the 2010 ACM Conference on Human Factors in Computing Systems.* New York, NY: ACM.

Choi, S. M., Kim, Y., Sung, Y., & Sohn, D. (2011). Bridging or Bonding? A cross-cultural study of social relationships in social networking sites. *Information Communication and Society, 14*(1), 107–129. doi:10.1080/13691181003792624

Chua, S.-C., & Choib, S. M. (2010). Social capital and self-presentation on social networking sites: A comparative study of Chinese and American young generations. *Chinese Journal of Communication, 3*(4), 402–420. doi:10.1080/17544750.2010.516575

Colás, P., & ... . (2013). Young People and Social Networks: Motivations and Preferred Use. *Scientific Journal of Media Education, 2*(40), 15–23.

Coleman, J. S. (1988). Social Capital in the Creation of Human Capital. *American Journal of Sociology, 94*, 95–121. doi:10.1086/228943

Croucher, S. M. (2011). Social Networking and Cultural Adaptation: A Theoretical Model. *Journal of International and Intercultural Communication, 4*(4), 259–264. doi:10.1080/17513057.2011.598046

Donath, J., & Boyd, D. (2004). Public Displays of Connection. *BT Technology Journal, 22*(4), 71–82. doi:10.1023/B:BTTJ.0000047585.06264.cc

Dunne, A., Lawlor, M. A., & Rowley, J. (2010). Young People's Use of Online Social Networking Sites: A Uses and Gratifications Perspective. *Journal of Research in Interactive Marketing, 4*(1), 46–58. doi:10.1108/17505931011033551

Ellison, N. (2008). Introduction: Reshaping campus communication and community through social network sites. In The ECAR study of undergraduate students and information technology (vol. 8, pp. 19–32). Boulder, CO: Educause.

Ellison, N., Seinfield, C., & Lampe, C. (2006). *Spatially Bounded Online Social Networks and Social Capital: The Role of Facebook*. Paper presented at the Annual Conference of the International Communication Association (ICA), Dresden, Germany.

Ellison, N., Seinfield, C., & Lampe, C. (2007). The Benefits of Facebook "Friends": Social Capital and College Students' Use of Online Social Network Sites. *Journal of Computer-Mediated Communication, 12*(4), 1143–1168. doi:10.1111/j.1083-6101.2007.00367.x

Ellison, N., Seinfield, C., & Lampe, C. (2011). Connection Strategies: Social Capital Implications of Facebook-Enabled Communication Practices. *New Media & Society, 13*(6), 873–892. doi:10.1177/1461444810385389

Ellison, N., Vitak, J., Gray, R., & Lampe, C. (2014). Cultivating Social Resources on Social Network Sites: Facebook Relationship Maintenance Behaviors and Their Role in Social Capital Processes. *Journal of Computer-Mediated Communication, 19*(4), 855–870. doi:10.1111/jcc4.12078

Fukuyama, F. (2001). Social Capital, Civil Society and Development. *Third World Quarterly, 22*(1), 7–20. doi:10.1080/713701144

Gershuny, J. (2002). Social Leisure and Home IT: A Panel Time-Diary Approach. *IT & Society, 1*, 54–72.

Granovetter, M. (1973). The Strength of Weak Ties. *American Journal of Sociology, 78*(6), 1360–1380. doi:10.1086/225469

Greenhow, C., & Robelia, B. (2009). Old Communication, New Literacies: Social Network Sites as Social Learning Resources. *Journal of Computer-Mediated Communication*, *14*(4), 1130–1161. doi:10.1111/j.1083-6101.2009.01484.x

Hampton, K., & Wellman, B. (2003). Neighboring in Netville: How the Internet Supports Community and Social Capital in a Wired Suburb. *City & Community*, *2*(4), 277–311. doi:10.1046/j.1535-6841.2003.00057.x

Haythornthwaite, C. (2000). Online Personal Networks. *New Media & Society*, *2*(2), 195–226. doi:10.1177/14614440022225779

Hofer, M., & Aubert, V. (2013). Perceived Bonding and Bridging Social Capital on Twitter: Differentiating between Followers and Followees. *Computers in Human Behavior*, *29*(6), 2134–2142. doi:10.1016/j.chb.2013.04.038

Islam, K. M (2006). Social Capital and Health: Does Egalitarianism Matter?: A Literature Review. *International Journal for Equity in Health*, *5*(3), 1–28. PMID:16597324

Johnston, K. (2013). Social Capital: The Benefit of Facebook 'Friends'. *Información Tecnológica*, *32*(1), 24–36.

Karuppasamy, G., Anwar, A., Bhartiya, A., Sajjad, S., Rashid, M., Mathew, E., ... Sreedharan, J. (2013). Use of Social Networking Sites among University Students in Ajman, United Arab Emirates. *Nepal Journal of Epidemiology.*, *3*(2), 245–250. doi:10.3126/nje.v3i2.8512

Kavanaugh, A., Carroll, J. M., Rosson, M. B., Zin, T. T., & Reese, D. (2005). Community Networks: Where offline Communities Meet Online. *Journal of Computer-Mediated Communication*, *10*(4). doi:10.1111/j.1083-6101.2005. tb00266.x

Kavanaugh, A., & Patterson, S. (2001). The Impact of Community Computer Networks on Social Capital and community involvement. *The American Behavioral Scientist*, *45*(3), 496–509. doi:10.1177/00027640121957312

Kohut, A. (2012). *Social Networking Popular Across Globe: Arab Publics Most Likely to Express Political Views Online*. Global Attitude Project, PEW Research Center. Retrieved from http://www.pewglobal.org/files/2012/12/Pew-Global-Attitudes-Project-Technology-Report-FINAL-December-12-2012.pdf

Koku, E., Nazer, N., & Wellman, B. (2001). Netting Scholars: Online and Offline. *The American Behavioral Scientist*, *44*(10), 1752–1774. doi:10.1177/00027640121958023

Lee, S. J. (2009). Online Communication and Adolescent Social Ties: Who benefits more from Internet Use? *Journal of Computer-Mediated Communication, 14*(3), 509–531. doi:10.1111/j.1083-6101.2009.01451.x

Lenhart, A., & Madden, M. (2007). *The Use of Social Media Gains a Greater Foothold in Teen Life*. Pew Internet and American Life Project.

McQuail, D., Blumler, J. G., & Brown, J. (1972). The television audience: A revised perspective. In D. McQuail (Ed.), *Sociology of Mass Communication* (pp. 65–135). Middlesex, UK: Penguin.

Mourtada, R., & Salem, F. (2011). Civil movements: The Impact of Facebook and Twitter. *Arab Social Media Report, 1*(2), 1–30.

Nie, N. H. (2001). Sociability, Interpersonal Relations, and the Internet: Reconciling Conflicting Findings. *The American Behavioral Scientist, 45*(3), 420–435. doi:10.1177/00027640121957277

Pempek, T. A., Yermolayeva, Y. A., & Calvert, S. L. (2009). College students' social networking experiences on Facebook. *Journal of Applied Developmental Psychology, 30*(3), 227–238. doi:10.1016/j.appdev.2008.12.010

Phua, J. J., & Jin, S. A. (2011). 'Finding a Home Away From Home': The Use of Social Networking Sites by Asia-Pacific Students in the United States for Bridging and Bonding Social Capital. *Asian Journal of Communication, 21*(5), 504–519. do i:10.1080/01292986.2011.587015

Putnam, R. D. (2000). *Bowling Alone: The Collapse and Revival of American Community*. New York: Simon & Schuster. doi:10.1145/358916.361990

Raacke, J., & Bonds-Raacke, J. (2008). MySpace and Facebook: Applying Uses and Gratifications Theory to Exploring Friend Networking Sites. *CyperPsychology & Behavior, 11*(2), 169–174. doi:10.1089/cpb.2007.0056 PMID:18422409

The Arab World Online Report: Trends in Internet Usage in the Arab Region. (2013). Dubai School of Government.

The State of Broadband 2014: Broadband for all. (2014). A report by the Broadband Commission, United Nations: Educational, Scientific, Cultural Organization.

Tice, D. M., Butler, J. L., Muraven, M. B., & Stillwell, A. M. (1995). When Modesty Prevails: Differential Favorability of Self-Presentation to Friends and Strangers. *Journal of Personality and Social Psychology, 69*(6), 1120–1138. doi:10.1037/0022-3514.69.6.1120

United Arab Emirates' National Bureau of Statistics. (2010). Retrieved on 24th January 2015, from: http://www.uaestatistics.gov.ae/ReportPDF/Population%20 Estimates%202006%20-%202010.pdf

Valenzuela, S., Park, N., & Kee, K. F. (2009). Is There Social Capital in a Social Network Site?: Facebook Use and College Students' Life Satisfaction, Trust, and Participation. *Journal of Computer-Mediated Communication*, *14*(4), 875–901. doi:10.1111/j.1083-6101.2009.01474.x

Villegas, J. (2011). The Influence of Social Media Usage and Online Social Capital on Advertising Perception. *American Academy of Advertising Conference Proceedings*, 69.

Vitak, J. (2012). The Impact of Context Collapse and Privacy on Social Network Site Disclosures. *Journal of Broadcasting & Electronic Media*, *56*(4), 451–470. do i:10.1080/08838151.2012.732140

Walkins, S. C., & Lee, H. E. (2009). *Bonding, Bridging and Friending: Investigating the Social Aspects of Social Networking Sites.* Paper presented in the 95th Annual Convention of the National Communication Association (NCA), Chicago. IL.

Wellman, B., Haase, A. Q., Witte, J., & Hampton, K. (2010). Does the Internet Increase, Decrease or Supplement Social Capital? Social Networks, Participation and Community Commitment. *The American Behavioral Scientist*, *45*(3), 436–455. doi:10.1177/00027640121957286

Wiest, J., & Eltantawy, N. (2012). Social media use among UAE College Students One Year after the Arab Spring. *Journal of Arab and Muslim Media Research*, *5*(3), 209–226. doi:10.1386/jammr.5.3.209_1

Williams, D. (2006). On and Off the 'Net: Scales for Social Capital in an Online Era. *Journal of Computer-Mediated Communication*, *11*(2), 593–628. doi:10.1111/j.1083-6101.2006.00029.x

Wong, W. K. W. (2012). Faces on Facebook: A Study of Self-Presentation and Social Support on Facebook. *Discovery-SS Student E-Journal*, *1*, 184–214.

Chapter 5

Users Holding Accounts on Multiple Online Social Networks:
An Extended Conceptual Model of the Portable User Profile

Sarah Bouraga
University of Namur, Belgium

Ivan Jureta
Fonds de la Recherche Scientifique, Belgium & University of Namur, Belgium

Stéphane Faulkner
University of Namur, Belgium

ABSTRACT

The last decade has seen an increasing number of online social network (OSN) users. As they grew more and more popular over the years, OSNs became also more and more profitable. Indeed, users share a considerable amount of personal information on these sites, both intentionally and unintentionally. And thanks to this enormous user base, social networks are able to generate recommendations, attract numerous advertisers, and sell data to companies. This situation has sparked a lot of interest in the research community. Indeed, users grow more uncomfortable with the idea that they do not have full control over their own data. The lack of control can even be amplified when a user holds an account on various OSNs. The data she shares is then spread over multiple platforms. This chapter addresses the notion of portable profile, which could help users to gain more control or more awareness of the data collected about her. In this chapter, the authors discuss the advantages and drawbacks of a portable profile. Secondly, they propose a conceptual model for the data in this unified profile.

DOI: 10.4018/978-1-5225-5715-9.ch005

INTRODUCTION

An area of the Web 2.0 gaining increasing success globally is the Online Social Network (OSN), or Social Networking Site (SNS). OSN refers to, according to Ellison et al. (2007):

Web-based services that allow individuals to (1) construct a public or semi-public profile within a bounded system, (2) articulate a list of other users with whom they share a connection, and (3) view and traverse their list of connections and those made by others within the system. The nature and nomenclature of these connections may vary from site to site."

The last decade has seen an increasing number of OSN users. These systems allow their users to interact with one another. Users set up an account, state relationships with other users, and are then able to communicate with each other, and share content. The most popular OSNs, such as Facebook, Twitter, or LinkedIn count hundreds of millions of members, that is, of users who have registered and thereby can use the features of these OSNs.

The first social network site was introduced in 1997. Called SixDegrees.com, it allowed its members to create a profile, list their Friends and, later to view others' friends lists (Ellison et al., 2007). As they grew more and more popular over the years, OSNs became also more and more profitable. Users share a considerable amount of personal information on these sites, both intentionally and unintentionally. And thanks to this enormous user base, OSNs are able to generate recommendations; attract numerous advertisers; and sell data to interested third parties.

This situation has led to the users' growing reluctance to share information. Users are uncomfortable with the idea that they do not have full control over their own data. In response, most current OSNs offer the possibility to their members to manage their privacy settings; allowing them to control who sees what about them and the content that they shared.

The increasing popularity of the OSN, and the questionable use of the data by the OSN have led to considerable interest in the research community. Many authors, for instance, have addressed the privacy and trust issues (Dwyer et al., 2007; Strater & Lipford, 2008; Madejski et al., 2011).

A way to increase the user trust in OSNs could be the introduction of a portable profile. It would be portable, in the sense that if a user registers on OSN A, she would be able to carry over the content of her profile to OSN B, and choose which of that data and content would appear on OSN B. The portable profile would offer more transparency to users, as they would know what data in some sense defines them on an OSN. This topic about the introduction of an integrated profile has also been mentioned in the literature (Heckmann et al., 2005; Berkovsky et al., 2008; Abel et al., 2011; Kapsammer et al., 2012).

This paper has two objectives and corresponding contributions. Firstly, we aim at listing the advantages and drawbacks of a portable profile, from the perspective of the user and from the perspective of the OSN. We will identify these benefits and limitations via an example. Secondly, we find the content for the portable user profile by looking at the content of user profiles on various existing OSNs, and from there propose a preliminary conceptual model of the portable user profile.

The rest of this article is organized as follows. Related work is introduced in Section 2. In Section 3, we discuss the motivations for a Portable User Profile (PUP). The proposed conceptual model for PUP is presented in Section 4. Finally, we discuss the results and conclude the paper in Sections 5 and 6 respectively.

LITERATURE REVIEW

Online Social Networks

Various aspects of OSNs have been studied. For example, there are studies focusing on the properties of the graph induced by connections between users. Some of the analyzed OSNs include: Flickr, YouTube, LiveJournal and Orkut (Mislove et al., 2007); MySpace and Orkut (Ahn et al., 2007); Sina blogs and Xiaonei SNS, two large Chinese online social networks (Fu et al., 2008); Flickr and Yahoo! 360 (Kumar et al., 2010); and Twitter (Kwak et al., 2010).. For an example of findings, consider Mislove et al. (2008), who found that various OSNs, despite their different purposes, share a number of similar structural features, namely: highly skewed degree distribution, a small diameter, and significant local.

The reasons why people want to use an OSN have also been examined in the literature. Various authors have studied the reasons motivating teenagers (Livingstone, 2008), college students (Park et al., 2009) or young adults (Subrahmanyam et al., 2008) to use an OSN. These reasons range from satisfying a *"friend"* and connection needs to having an additional source of information (Raacke & Bonds-Raacke, 2008; Subrahmanyam et al., 2008).

The privacy issue was examined by several authors. They include the study of privacy on Facebook (Strater & Lipford, 2008; Madejski et al., 2011); as well as the comparison of trust and privacy issues on Facebook and MySpace (Dwyer et al., 2007).

Specific social networks were also studied in more details: MySpace (Caverlee &Webb, 2008); Facebook (Lampe et al., 2006; Pempek et al., 2009); YouTube (Lange, 2007); and Massively Multiplayer Online Games (MMOGs) (Ducheneaut et al., 2006).

User Profile

In the area of OSNs, the topic of the user profile has been researched. Some authors have proposed approaches to compare the profiles of two users, more specifically to measure the distance between two user profiles in an OSN (Rezaee et al., 2012). The similarity between users has been measured based on the information on their Orkut profile (Singh & Tomar, 2009); the correlation between the similarity of two aNobbi (an OSN for book lovers) users' profile and the link between these two users (Aiello et al., 2010). Mislove et al. (2010) also explored the possibility to infer the attributes of some users in an OSN, given the attributes of some users in the same OSN. Our work here is different from these because we are trying to define a generic profile, that would fit any type of OSNs. The cited works here compare user profiles based on their content. We are trying to identify this content before being able to compare two profiles or to execute any kind of manipulations on the user profile.

Other authors have carried out sociological studies based on the user profile. Some researchers studied the popularity of users. More specifically, Lampe et al. (2007) studied the role played by elements of a Facebook user profile in the creation of online connections. They discovered that the more fields a user populates in her profile; the more friends she will have. The factors determining the profile popularity in a professional social network were also examined (Strufe, 2010). Utz (2010) studied how the perceived popularity, communal orientation, and social attractiveness of a user on an OSN (here Hyves) are influenced by the user's extraversion, the extraversion of the user's friends, and the number of friends a user has. Utz et al. (2011) examined the effects of OSN use on romantic relationships. Dunbar et al. (2015) used Facebook and Twitter profiles to create ego-centric social networks. The authors discovered that the structure of OSN mirrors those in the face-to-face networks. Our work here is different from these because we are not interested in the sociological aspect of the user profile. Sociological aspects can have an influence and we investigated that in another work (Bouraga et al., 2015), where we identified factors that can have an influence on the perceived relevance of content. However, for the definition of a generic profile, we are not interested in that area.

The type of information users share on their profile was evaluated by several authors: Nosko et al. (2010) examined the kinds of information Facebook users shared on their profile; Emmanuel et al. (2013) examined the type of information users share on their profile, depending on the context of the social network (the dating OSN and the professional OSN). Silfverberg et al. (2011) explored the effort members of Last.fm (an OSN dedicated to music preferences) invest in the process of maintaining their profile (they called this process "profile work"); that is in the process of controlling their "self-presentation". Chen et al. (2014) modeled profile privacy settings from a game theoretic perspective. Their model shows that users choose for the highest possible privacy if they encounter any risk, regardless of any incentive for profile disclosure. Our work presents some similarities with the cited works here, because the latter identify the types of information shared by users on their profile. However, the difference between the existing work and our work lies in the generality of the models. The cited works define the types of information in a specific context (the information shared on the Facebook profile, the information shared depending on the context of the OSN, the information shared in the context of self-presentation, and in the context of privacy settings management). However, in this paper, we aim to identify generic attributes of a user profile.

Various authors examined the link between the information users share on their profile and the threats they can face as an OSN user. Given the amount of personal information shared on OSN, users are vulnerable to "social engineering attacks". Alim et al. (2011) proposed an automated approach to extract profile data in order to assess the vulnerability of the user. Kontaxis et al. (2011) developed a tool to automatically detect social network profile cloning. Also, My3, a "privacy- aware decentralized OSN", was proposed, enabling its users to have "full access control on their data" (Narendula et al., 2011). Fire et al. (2014) reviewed various threats OSN users can face, such as fake profile or face recognition, and proposed solutions to address these threats, such as fake or cloned profile detection. Before conducting further analysis, this paper aims at the definition of a generic profile. Thus, we are not interested, for now, in the threats a user faces when sharing certain types of information.

Authors are aware that people hold an account on different OSNs. Therefrom, various solutions were proposed to identify the same user on different OSNs. Examples include Nie et al. (2016) who introduced the Dynamic Core Interests Mapping (DCIM) algorithm to match the same user on various OSNs; and Ma et al. (2017) who proposed a solution called MapMe based on both the user profile as well as the user relationship network structures. Komamizu et al. (2017) identified the same users who were both on Github and Stackoverflow. And Zhou et al. (2016) proposed the FRUI algorithm based on friend relationship. The purpose of this article is different from these studies because we are not interested in identifying the same user with a different OSN profiles, rather we want to propose a conceptual model that will gather all the information shared by the user on these various OSNs.

Finally, several researchers have examined the interest of integrating data from various OSN profiles into one user profile. Zhang et al. (2014a, 2014b) addressed the linkage of people having a profile on different OSNs. The difficulty lies in the fact that people may have different usernames on different OSNs. The difference between their work and ours is the motivation behind it. Zhang et al. aimed for a holistic understanding of the OSN user; while we aim to propose a conceptual model of a portable user profile. Heckmann et al. (2005) introduced "GUMO - the General User Model Ontology". The authors used OWL as ontology language to represent "the user model terms and their interrelationships". The motivation for their work is similar to ours, that is, the authors sought "the simplification for exchanging user model data between different user-adaptive systems". However, our work is different from GUMO with regard to two elements. First, we focus here on the OSNs, while GUMO has a larger scope and is meant for intelligent semantic web environments. Secondly, Heckmann et al. developed their ontology for the benefits of the systems; while we take into account the user perspective. Berkovsky et al. (2008) proposed an approach to import and integrate data from several remote RSs to enhance the efficiency of another RS. The recommendations are more accurate thanks to more complete user profiles. The authors call the process of importing and integrating data "the mediating process". More specifically, they define the latter as "mediation of user models is a process of importing the user modeling data collected by other (remote) recommender systems, integrating them and generating an integrated user model for a specific goal within a specific context" (Berkovsky et al., 2008). Abel et al. (2011) examined cross-system user modeling strategies, consisting of the following building blocks: source of user data, semantic enrichment, and weighting scheme. The authors, then, evaluate the strategies' "performance in generating valuable profiles in the context of tag and resource recommendation in Flickr, Twitter, and Delicious". Kapsammer et al. (2012) proposed a "semi-automatic approach to derive social network schemas from social network data". Their process consists of four phases, namely: data extraction, schema extraction, transformation, and integration.

MOTIVATION FOR CENTRALIZED PROFILE

Throughout this Section, we will look at a hypothetical user who is using Facebook, LinkedIn, and Pinterest. We chose these three OSNs because they have different purposes, and the PUP should not be specific to the kind of OSN. Facebook and LinkedIn are network-oriented, that is they put the emphasis on the relationships between users. However, they still differ in purposes. The former is used by people who want to connect with friends; while the latter is used for business contacts.

Pinterest is of the knowledge-sharing type, that is the focus is on the content sharing, instead of the relationships (Guo et al., 2009).

To use all three OSNs under the same offline identity, the user has to fill in the same information at all three OSNs, such as name, email, location, gender, profile picture, an "About you" section, and so on. Facebook and LinkedIn also offer the user to share the following elements: her occupation (the school she attends or attended), the job(s) she has (had); her skills (Facebook communicates the languages the user can speak, while LinkedIn introduces the users qualifications, summary, areas of expertise); her interests.

On all three OSNs, the user has to find friends and/or other users to follow. The more active a user is on a social network, the more she can get out of it. The network effect is that the more friends you have, the more valuable the experience can be.

Currently, business models of OSNs aim to make profit by (i) allowing companies to post targeted ads, based on user data; and (ii) selling data about usage to companies. It follows that OSNs do not have incentives to integrate their user data and provide the users with a portable and accessible profile. On the contrary, OSNs are incentivized to keep all the data they gather about their users for themselves, since these data constitute the product they sell to advertisers. Several OSNs allow some connections with other OSNs, to sign in, to share post, and/or find friends. For instance, users of Facebook and LinkedIn can share post across networks, that is they can post simultaneously on both OSNs; also Facebook users can sign in and find friends more easily on Pinterest with their Facebook account. However, this form of collaboration already impairs the collection of user data by social networks; as evidenced by Yahoos decision in March 2014 to discard the use of Facebook and Google to sign in and log into Yahoo services.

Benefits of a Portable User Profile

Several authors in the literature have already put forward some benefits of an "integrated" profile, such as: an enriched user profile (Abel et al., 2010), and improved quality of the recommendations (Berkovsky et al., 2008; Abel et al., 2011; Kapsammer et al., 2012).

Abel et al. (2010) argue that a profile aggregation can lead to an enriched user profile. An aggregated profile offers "significantly more information about the users than individual service profiles can provide". Profile aggregation can be used to improve incomplete profiles. Thus, the user could fill in her profile only once but the information will still be present in every site she uses.

Example 1: Without the PUP: The user has to enter all her personal information on every OSN that she wants to use, here Facebook, Pinterest, and LinkedIn. All three OSNs require an email address and a profile picture. Facebook and Pinterest ask for the name of the user, and offer an "About you" space. Without an integrated or portable profile, the user has to repeatedly provide the same information. Also, the user has to create links on the three OSNs. If she has friends who use the same OSNs as her, she has to create relationships with them on all three OSNs.

Example 2: With the PUP: With a portable profile, the user could share her personal information only once, whether it is on Facebook, Pinterest, or LinkedIn; and this information will be reused by the other OSNs. The user could fill in several fields in her Facebook profile, and this information could then be used by LinkedIn or Pinterest to complete the same users profile. As far as the connections between users are concerned, it would be easier to find friends on different OSNs. For instance, assume that the user is friends with A on Facebook. A decides to set up an account on LinkedIn. Because all the information about the user is stored in an integrated profile, including the links the user has on all the OSNs, the user and A could automatically be linked on LinkedIn. This assumes that the PUP includes the data on relationships the user has on other OSNs.

The PUP leads to another advantage, namely the improved quality of the recommendations (Berkovsky et al., 2008; Abel et al., 2011; Kapsammer et al., 2012).

Because of the enhanced information, the cold start problem is avoided; that is the recommendation algorithms could avoid suffering from data sparsity (Adomavicius & Tuzhilin, 2005). The most common recommendation techniques rely on a large amount of data to generate recommendations. The accuracy of the recommendation depends, thus, on the given data. If the user is new and/or is not active, the recommendations she gets will most likely be of poor quality.

Example 3: Without the PUP: The accuracy of the recommendation varies across OSN. Depending on the activity of the user, the recommendation will be more or less accurate. If the user is very active on Facebook, the recommendations she will get will likely be of high quality. The OSN will make friend suggestions, or will suggest pages to like that correspond to the user profile. However, if the user barely spends any time on Pinterest, then the OSN will probably struggle with the generation of qualitative "pin" (post on Pinterest) or "pinner" (user of Pinterest) recommendations.

Example 4: With the PUP: The accuracy of the recommendation is stable across OSNs. The quality of the recommendations will not depend on the activity of the user on the particular OSN. The user could be more active on Facebook, but it will not impair the quality of the recommendations generated by Pinterest or LinkedIn. The Facebook activity will be stored in the integrated profile and the other OSNs will take advantage of this information to produce recommendations to that user. Furthermore, even if the user were equally active on Facebook and Pinterest, the information gathered by the OSNs would be different. Hence, a portable profile would allow OSNs to have complementary information at their disposal.

A portable profile could also directly benefit the user. The latter would have access to the data OSNs have gathered about her, both explicit data (that is, data the user has explicitly shared with the OSN), and implicit data (that is, data about the user activity). The user could then control and manage the elements present in her integrated profile. This situation could help users accept the recommendations she gets, and it could make the suggestions of items less intrusive.

Example 5: Without the PUP: The user only has control over the data she gives explicitly. The user manages what she posts on Facebook, what she shares on LinkedIn, and what she "pins" on Pinterest. The user can also control the links she creates on all these OSNs. However, the user cannot exactly know what the OSNs gather about her. The user cannot know where the recommendations she gets come from; which could render the latter a little bit intrusive.

Example 6: With the PUP: The user can have control over the data she gives explicitly and the data gathered by the OSNs. The user can manage her activity on Facebook, and can decide if this activity can be used by LinkedIn. The user can figure out why she was recommended a particular job on LinkedIn, or why she was given this particular friend suggestion on Facebook.

Limitations of a Portable User Profile

Abel et al. (2010) identified a "risk of intertwining user profiles, namely that users who deliberately leave out some fields when filling their Twitter profile might not be aware that the corresponding information can be gathered from other sources". A unified profile implies unified information across OSN. Users cannot decide to share more on a OSN, and cannot keep information from being used by a specific OSN.

Example 7: Without the PUP: The user can decide to share more information on a particular OSN. She can decide to be more active on Facebook, for instance; and she does not want LinkedIn or Pinterest to have as many information at their disposal. She can use different profile pictures, or usernames for each OSN. Also, she can decide on the links she wants to create depending on the OSN. She can agree to a friend request on Facebook from a colleague, but she can decide not to share a Pinterest account with that colleague.

Example 8: With the PUP: If the user cannot control which part of the PUP is carried over between OSNs, then she cannot choose which information she wants to share on which OSN. The user has to be more careful about the information she posts on Facebook, because that information can then be used by LinkedIn or Pinterest. For instance, if the user accepts the friend request on Facebook from her colleague; then, if they are both on Pinterest, the link could be automatically created in that OSN. The user cannot separate or compartmentalize her online activity.

Another important issue related to the portable profile is related to privacy. Users post large amount of personal information online. But in a way, the user has more control over what is known about her by each OSN, when she has separate profiles. Indeed, with a portable profile, more information can be discovered by the OSNs. Inferences could be made more accurate because they are based on data coming from various sources. The user is thus more "vulnerable to social engineering attacks" (Alim et al., 2011).

Example 9: Without the PUP: As mentioned above, the user can decide which information she posts on which OSN. We will take the example of the horoscope and the age: "if the age and horoscope signs are present on a profile then it is possible to guess when the birthday is" (Alim et al., 2011). The user can have a board on Pinterest with her horoscope sign, and she can mention her age on Facebook. But the user can decide to keep her exact birthday private. Because the information about her sign, and her age are mentioned on different OSNs, it will be difficult to infer her birth date.

Example 10: With the PUP: The user is more vulnerable; she has less control over what can be known/ discovered about her. Indeed, even if she posts her age only on Facebook, and her horoscope sign on Pinterest; that information will be present on the integrated profile. The latter can then be used to infer more personal information about the user, and in our example her exact birthday can be discovered.

PORTABLE USER PROFILE CONCEPTUAL MODEL

Data Used in Facebook, LinkedIn, and Pinterest

Before proposing the PUP conceptual model, we will identify the information that the user gives to the OSNs considered here, namely Facebook, LinkedIn, and Pinterest. We can distinguish three categories of data used by OSNs: the profile data, the relationships information, and the posts. The profile category includes the following pieces of information: her login information, her identity, her occupation, her beliefs, information about her family, her skills, her interests, and a text about herself. Then, the relationships the user creates with other users can also be part of the user profile. These relationships can be categorized in one of two groups: unidirectional relationships, or bidirectional relationships. The former type of relationships is unreciprocated, that is a user likes, or subscribes to a fan page; or follows another user. The latter type of relationships is reciprocated. A friend request is sent by a user to another user, who has to accept or deny the friend request. If she confirms it, then the relationship is created. Otherwise, no link exists between the two users. The third category of profile data is the posts shared by the user, that is the texts, comments, like/repost, tags, media, messages and groups. Those data are not directly and consciously given by the user; rather the OSN gathers this information for every user.

Facebook allows its users to login with their email or phone. It asks the users about their name, birthdate, gender, address, phone number, school, job, family, religious and political views, languages they can speak, favorite quotation, interests, and asks for a profile picture; Facebook also offers an "About you" section. The OSN supports both types of relationships. The user can send and accept/ deny a friend request; and she can also like a page about an artist, a public person, or a company. Facebook supports various types of posts, and hence gathers information about the following user activity: status, notes and links posted by the user; comments on status, or media; like/share; tag of friends on status or media; photos and videos; messages; and groups created or joined by the user.

LinkedIn users can login with their email address. The OSN asks them about their name, title, location, and a profile picture. The OSN also asks about the industry the user works in, and her experience, her qualifications, summary, specialties, specific skills and areas of expertise; her interests; and her personal details and advice for contacting her. LinkedIn supports both types of relationships. Users can ask other users to connect with them; and users can also follow companies' page. LinkedIn can gather information about the activity of the users, that is, their summary, recommendations; comments; like/share; the connections they tag on status: photos and videos; message; and groups.

Pinterest allows their users to login with their email address or their Facebook, Google, or Twitter account. Pinterest asks for less information than the other two OSNs. Indeed, pinners are asked to provide their name, location, website, gender, and profile picture; and they can also fill in an "About you" section. The relationships in Pinterest are unidirectional. Users follow other users; they do not have to send a friend request to have access to other user's profiles. Pinterest gathers information about the pins posted by users; the comments; the like/repin; tag; photos and videos; messages; and group boards.

This discussion is summarized in Tables 1 and 2.

Structure of the Data

The user profile is composed of two main categories of data: "Explicit" and "Implicit" data. The first category, the explicit data are the data the user intentionally and directly gives the OSN. The user is fully aware that she gives away that information. That category can be further classified in two groups: the "Profile Information" data and the "Relationships Information" data. The former group consists of: the login information; the identity of the user; her occupation; her beliefs; her skills; her hobbies; and information taking the form of an "About you" section. The latter group, that is the relationships data, consists of information about the links the user

Table 1. Data Used in Facebook, LinkedIn, and Pinterest - Part 1

	Login	Identity	Occupation	Family/ Beliefs
Facebook	Email, phone	Demographics, picture	School(s), Job(s)	Family, Religious, Political
LinkedIn	Email	Demographics, picture	Industry, Experience	/
Pinterest	Email	Demographics, picture	/	/

Table 2. Data Used in Facebook, LinkedIn, and Pinterest - Part 2

	Skills	Interests	Else	Relationships	Post
Facebook	Languages	V	About me	Bidirectional and Unidirectional	V
LinkedIn	Summary, Expertise	V	Personal details, Contact me	Bidirectional and Unidirectional	V
Pinterest	/	/	About you	Unidirectional	V

creates with other users. In that group, we can find both the bidirectional relationships and the unidirectional relationships.

Thus, two subclasses belong to the class "Explicit" data: Profile and Relationships. Both subclasses can be further specialized. For each subclass, we will identify: (i) the classes specializing the subclass; (ii) the attributes of these classes with their cardinality and their type; and (ii) the potential constraint(s).

We will start by the Profile Information subclass:

- Login information
  - Attributes
    - Email address [0..1], string
    - Phone number [0..1], number
    - Password [1..1], string
  - Constraint
    - The user has at least one value for the email address or the phone number
- Identity, specialized into
  - Identification – Attributes:
- First name [1..1], string
- Last name [1..1], string
- Birthday [0..1], date
- Gender [0..1], string
- Mother language [0..1], string
- Ethnicity [0..1], string
  - Location – Attributes:
- Street [0..1], string
- Home number [0..1], string
- City [0..1], string
- Country [0..1], string
- Time zone [1..1], string
  - Relationship status – Attributes:
- Status [1..1], enumerate: Single, In a relationship, Engaged, Married, In a civil union, In a domestic partnership, In an open relationship, It's complicated, Separated, Divorced, Widowed
- Start date [1..1], date
- End date [0..1], date
- Partner ID [0..1], integer
  - Phone number – Attributes:
- Type [0..1], enumerate: Mobile, Home, Work, Fax
- Number [1..1], number

- ◦ Profile picture – Attributes:
- Photo ID [1..1], integer
- Size [1..1], integer
- ◦ Website – Attributes:
- URL [1..1], string
- Description [0..1], string
- ◦ Occupation information – Attributes:
- Type [0..1], enumerate: School, Job, Industry
- Name [1..1], string
- ◦ Family members – Attributes:
- ID of the family member [1..1], integer
- Relationship [1..1], string
- ◦ Beliefs information – Attributes:
- Type [0..1], enumerate: Religious, Political
- Belief [1..1], string
- ◦ Skills information – Attributes:
- Type [0..1], enumerate: Foreign languages, Qualifications/ Specialties, Area of expertise/Special skills
- Skill [1..1], string
- ◦ Hobbies information – Attributes:
- Type [0..1], enumerate: Interests, Kind of music, Kind of movies, Favorite quote
- Hobby [1..1], string
- ◦ Presentation information – Attributes:
- Type [1..1], enumerate: About you/Describe yourself/ Biography, Advice for contacting me
- Presentation [1..1], string

We will now turn to the Relationship class:

- Unidirectional
  - ◦ Attribute
    - ▪ ID of the followed user [1..1], integer
- Bidirectional
  - ◦ Attributes
    - ▪ ID of the friend [1..1], integer
    - ▪ Sender of the friend request [1..1], boolean

The second main category is the "Implicit" data, that is the data the user unintentionally gives to the OSN. Similar to the first category, the implicit data can be broken down into two classes, namely the "Posts", and the "Activity". In the posts, one can find all the elements shared by the user with her friends/contacts. This group contains the following elements: text, comment, like/repost, tag, media, message, and group. The activity group consists of the browsing and searching activity conducted by the user when she looks through the OSN.

Similar to the Explicit data, we will go through each concept of the Implicit class, and we will detail the attributes, the attributes' cardinalities, their type, and the potential constraints.

We will start with the Posts class:

- Text information
  - Attributes
    - Text ID [1..1], integer
    - Type [1..1], enumerate: Status, Moods, Quotes, Links, Notes
    - Text [1..1], string
- Comments information
  - Attributes
    - Comment ID [1..1], integer
    - Type [1..1], enumerate: On a profile information, On a text, On a media, On a relationship status
    - Comment [1..1], string
- Like/Repost information
  - Attributes
    - Like/Repost ID [1..1], integer
    - Type [1..1], enumerate: Like a text, Repost a text, Like a media, Repost a media
- Tag information
  - Attributes
    - Tag ID [1..1], integer
    - Type [1..1], enumerate: A friend on a media, A friend on a text, A media, A text
    - Tag [1..1], string
- Media information
  - Attributes
    - Media ID [1..1], integer
    - Type [1..1], enumerate: Photo, Video, Gif

- ▪ Name [1..1], string
- ▪ Size [1..1], integer
- Message information
  - ○ Attributes
    - ▪ Message ID [1..1], integer
    - ▪ Type [1..1], enumerate: Public message, Private message, Instant chat
    - ▪ Content [1..1], string
    - ▪ Sender ID [1..1], integer
    - ▪ Recipient ID [1..*], integer
  - ○ Constraint
    - ▪ A message can be commented, liked, or reblogged if it is of the type "Public message"
- Group information
  - ○ Attributes
    - ▪ Group ID [1..1], integer
    - ▪ Name of the group [1..1], string
    - ▪ Date joined [1..1], date
    - ▪ Creator of the group ID [1..1], integer

We will now turn to the Activity class:

- Browse information
  - ○ Attributes
    - ▪ Type [1..1], enumerate: User, Media, Text, Comment, Like/ Repost, Tag, Group, Message
    - ▪ Timestamp [1..1], date
- Search information
  - ○ Attributes
    - ▪ Search ID [1..1], integer
    - ▪ Keyword [1..*], string
    - ▪ Timestamp [1..1], date

We classified the Profile and the Relationships information in the Explicit category, and we classified the Posts information in the Implicit category for two reasons. Firstly, we believe that the user has to give her profile information, and has to create relationships online, create links with other users before using the OSN. She expects her information to be used by the OSN. However, the posts, and the activity represent the user actually using the OSN; the user may not be aware of the OSN using this type of information. The user shares posts with her friends, and not directly with the OSN. Secondly, today, when the concept of Profile is mentioned, people visualize only the Explicit class; while the OSN takes advantage of both the Explicit and the Implicit classes. By representing the conceptual model with both categories, we can give the user more control and more awareness over her data, that is over the data gathered by the OSN and not only the data explicitly given by the user. The Portable User Profile conceptual model is represented as a UML Class Diagram, in Figure 1.

DISUCSSION

Many OSN features allow the user to share content about herself, both directly and indirectly. Directly, by giving, for example, her birth date, marital status, and so on. Indirectly, by performing actions which suggest her preferences; for example, "liking" some specific content on Facebook gives indications about which topics and other users that user may be interested in. The availability of such, so to speak personal content, has led to concerns about, and research on trust and privacy (Dwyer et al., 2007; Strater et al., 2008; Madejski et al., 2011).

Figure 1. The Portable User Profile conceptual model

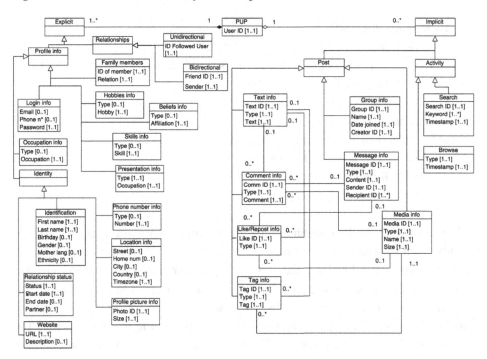

It is not possible, we believe, to study OSN features and content without paying particular attention to personal content. Why? Because we expect users to consider more important that personal content over other, and features for manipulating (sharing, hiding, removing, for example) that personal content over those focusing on other, impersonal content. In (Bouraga et al., 2015)], we studied the importance of content for OSN users, generated by other OSN users. More specifically, we address the following question: "How relevant is it to user X to know about some specific event type generated by user Y?". This question can be rephrased into: "How relevant is to user X to know about some specific content in the PUP of a user Y?"

In order to understand how personal content influences users' perception of relevance of other content and features on OSNs, in this paper we have proposed a definition of the personal content, in the form of a conceptual model of the Portable User Profile (PUP). The PUP of a user lists the types of data, and relationships between these types of data, which most well-known OSNs tend to collect. It is portable, in the sense that if it were possible for a user to carry over her personal content from one OSN to another, then the PUP includes data, which it would be relevant to carry over. Put another way, the PUP represents all the information that the user has shared across multiple OSNs.

Also, we believe that the conceptual model is scalable. It is not specific to one particular OSN nor a specific type of OSNs (for instance, for knowledge-sharing OSNs or network-oriented OSNs). If a new OSN arises, it should not be a problem to take into account the information it gathers, we should be able to find a mapping between its features and the proposed class diagram. And it can be easily extended if OSNs propose new features and therefrom, new types of personal content arise. More specifically, the PUP can be updated by adding a class to the conceptual model, accompanied by the potential relationships with other classes.

As mentioned in Section 3, the existence of a Portable User Profile has its benefits as well as its limitations. We will discuss the PUP conceptual model in light of these advantages and limitations.

Firstly, the PUP conceptual model enables the representation and the common understanding of the structure of the data/information in the user profile. The user can share her information only once, and this information will take the form of the conceptual model. Also, if the user is more active on a OSN, for instance, if she uses Facebook daily while she only checks her LinkedIn account once a week; the user can leverage her Facebook activity and take advantage of it on LinkedIn.

The distinction between the explicit and the implicit data is also made clear by the conceptual model: what is implied by "explicit data"? What is implied by "implicit data"? The user can have more control over her information, because she knows what is exactly tracked when she is on the OSN.

However, risks remain. What the user cannot control are the new inferences the OSNs can make (which are not represented in the PUP). The user can identify where the information come from, but cannot control the information that is leveraged from all the various sources. Also, the user cannot choose to share more information on an OSN, and less on another. She cannot hide data from one OSN. The information at the disposal of the OSN is uniform. The PUP conceptual model does not erase the limitations of an integrated profile, but it makes the user more aware of those risks.

We will now discuss the recommendation topic, in light of the PUP conceptual model. The quality of the recommendations depends on the information it is based on, both in terms of quality and quantity.

Adomavicius & Tuzhilin (2005) distinguish between three recommendation techniques that a Recommendation System (RS) can use, each of which present several shortcomings:

1. Collaborative Filtering (CF), where "the user is recommended items that people with similar tastes and preferences liked in the past";
    a.   New user problem: the RS has to learn the user's preferences in order to make reliable recommendations
    b.   New item problem: an item has to be rated by a significant amount of users before it can be recommended
    c.   The "grey sheep" problem: a user can be classified in more than one group of users
    d.   Sparsity: both users and ratings sparsity can cause problems for the generation of accurate recommendations
2. Content-Based (CB), where "the user is recommended items similar to the ones the user preferred in the past";
    a.   Limited content analysis: a sufficient set of features per item is required in order to produce recommendations
    b.   Over-specialization: the set of recommended items will be very homogeneous
    c.   New user problem
3. Hybrid, which is a combination of CF and CB.

A RS could also generate recommendations based on rules. In (Bouraga et al., 2015). we proposed rules for relevant recommendations. We surveyed students of the University of Namur, and based on their preferences, we proposed decision trees for relevant recommendations.

Thus, the more information the RS can use, the more accurate it will be; whether the RS uses CF, a CB, a hybrid, or rules. The information has to be varied as well, and it has to be of good quality. The recommendation will be the most accurate if the user shares all the classes of information found in the PUP conceptual model. More specifically, if she shares her profile information, if she creates relationships online, and if she is active on the OSN that is, if the OSN can gather lots of implicit data. To the contrary, if the user shares only the minimum amount of data, that is her profile information, she will most likely get imprecise recommendations. This discussion is represented in Figure 2. The quality of the recommendation increases as we move towards the external layers. The inner circle depicts the core information we believe is needed to generate recommendations; while the outer circle represents the optional information.

CONCLUSION

In this paper, we proposed the "Portable User Profile" (PUP) conceptual model. This model is meant to gather all the data/information a user shares on every OSN she uses. The PUP should be accessible by both the user, and the OSN.

Figure 2. The PUP conceptual model and the quality of recommendations

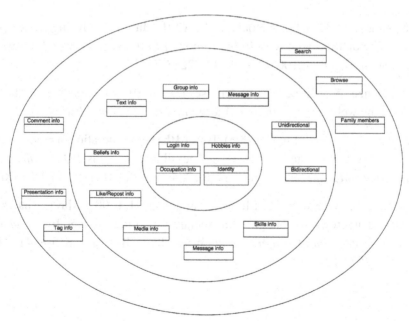

Before introducing the conceptual model, we first discussed the advantages and limitations of such a profile. An integrated profile is richer in terms of information than a single OSN account, and this enriched profile probably leads to better, more accurate recommendations. Also, the user could have access to all the information the OSNs gather about her; and thus the user could have more control over her own data. Nevertheless, with an integrated profile the user cannot choose an OSN where she would be more active, where she would share more information. She is forced to be equally exposed on every OSN she uses. This situation also renders the user more vulnerable. Our PUP conceptual model is composed of two main classes: Explicit and Implicit data. Each of these classes can be specialized into two classes: Profile and Relationship data; and Posts and Activity data respectively. Those four classes are specialized into several other classes. We represented this conceptual model as a UML Class diagram.

The main limitation of this work consists in its research-oriented/ hypothetical nature. We have not validated the proposal. Also, we only took into account three different OSNs; even though, these three different OSNs are different in their approach and objectives. Considering other OSNs would be interesting in order to validate or revise our PUP conceptual model. Another limitation is that we did not consider the information resulting from inference by the OSN or from the collection of information from other sources.

REFERENCES

Abel, F., Araújo, S., Gao, Q., & Houben, G. J. (2011, June). Analyzing cross-system user modeling on the social web. In *International Conference on Web Engineering* (pp. 28-43). Springer. 10.1007/978-3-642-22233-7_3

Abel, F., Henze, N., Herder, E., & Krause, D. (2010). Interweaving public user profiles on the web. *User Modeling, Adaptation, and Personalization*, 16-27.

Adomavicius, G., & Tuzhilin, A. (2005). Toward the next generation of recommender systems: A survey of the state-of-the-art and possible extensions. *IEEE Transactions on Knowledge and Data Engineering*, *17*(6), 734–749. doi:10.1109/TKDE.2005.99

Ahn, Y. Y., Han, S., Kwak, H., Moon, S., & Jeong, H. (2007, May). Analysis of topological characteristics of huge online social networking services. In *Proceedings of the 16th international conference on World Wide Web* (pp. 835-844). ACM. 10.1145/1242572.1242685

Aiello, L. M., Barrat, A., Cattuto, C., Ruffo, G., & Schifanella, R. (2010, August). Link creation and profile alignment in the aNobii social network. In *Social Computing (SocialCom), 2010 IEEE Second International Conference on* (pp. 249-256). IEEE. 10.1109/SocialCom.2010.42

Alim, S., Abdulrahman, R., Neagu, D., & Ridley, M. (2011). Online social network profile data extraction for vulnerability analysis. *International Journal of Internet Technology and Secured Transactions, 3*(2), 194–209. doi:10.1504/IJITST.2011.039778

Berkovsky, S., Kuflik, T., & Ricci, F. (2008). Mediation of user models for enhanced personalization in recommender systems. *User Modeling and User-Adapted Interaction, 18*(3), 245–286. doi:10.100711257-007-9042-9

Bouraga, S., Jureta, I., & Faulkner, S. (2015). An empirical study of notifications' importance for online social network users. *Social Network Analysis and Mining, 5*(1), 51. doi:10.100713278-015-0293-x

Bouraga, S., Jureta, I., & Faulkner, S. (2015), Rules for relevant notification recommendations to online social networks users. In *ASE Eighth International Conference on Social Computing (SocialCom)*. Academic Press.

Chen, J., Kiremire, A. R., Brust, M. R., & Phoha, V. V. (2014). Modeling online social network users' profile attribute disclosure behavior from a game theoretic perspective. *Computer Communications, 49*, 18–32. doi:10.1016/j.comcom.2014.05.001

Ducheneaut, N., Yee, N., Nickell, E., & Moore, R. J. (2006, April). Alone together?: exploring the social dynamics of massively multiplayer online games. In *Proceedings of the SIGCHI conference on Human Factors in computing systems* (pp. 407-416). ACM. 10.1145/1124772.1124834

Dunbar, R. I., Arnaboldi, V., Conti, M., & Passarella, A. (2015). The structure of online social networks mirrors those in the offline world. *Social Networks, 43*, 39–47. doi:10.1016/j.socnet.2015.04.005

Dwyer, C., Hiltz, S., & Passerini, K. (2007). Trust and privacy concern within social networking sites: A comparison of Facebook and MySpace. *AMCIS 2007 Proceedings*, 339.

Ellison, N. B. (2007). Social network sites: Definition, history, and scholarship. *Journal of Computer-Mediated Communication, 13*(1), 210–230. doi:10.1111/j.1083-6101.2007.00393.x

Emanuel, L., Bevan, C., & Hodges, D. (2013, April). What does your profile really say about you?: privacy warning systems and self-disclosure in online social network spaces. In CHI'13 Extended Abstracts on Human Factors in Computing Systems (pp. 799-804). ACM.

Fire, M., Goldschmidt, R., & Elovici, Y. (2014). Online social networks: Threats and solutions. *IEEE Communications Surveys and Tutorials*, *16*(4), 2019–2036. doi:10.1109/COMST.2014.2321628

Fu, F., Liu, L., & Wang, L. (2008). Empirical analysis of online social networks in the age of Web 2.0. *Physica A*, *387*(2), 675–684. doi:10.1016/j.physa.2007.10.006

Guo, L., Tan, E., Chen, S., Zhang, X., & Zhao, Y. E. (2009, June). Analyzing patterns of user content generation in online social networks. In *Proceedings of the 15th ACM SIGKDD international conference on Knowledge discovery and data mining* (pp. 369-378). ACM. 10.1145/1557019.1557064

Heckmann, D., Schwartz, T., Brandherm, B., Schmitz, M., & von Wilamowitz-Moellendorff, M. (2005). GUMO-the general user model ontology. Lecture Notes in Computer Science, 3538, 428.

Kontaxis, G., Polakis, I., Ioannidis, S., & Markatos, E. P. (2011, March). Detecting social network profile cloning. In *Pervasive Computing and Communications Workshops (PERCOM Workshops), 2011 IEEE International Conference on* (pp. 295-300). IEEE. 10.1109/PERCOMW.2011.5766886

Kumar, R., Novak, J., & Tomkins, A. (2010). Structure and evolution of online social networks. In *Link mining: models, algorithms, and applications* (pp. 337–357). Springer New York. doi:10.1007/978-1-4419-6515-8_13

Kwak, H., Lee, C., Park, H., & Moon, S. (2010, April). What is Twitter, a social network or a news media? In *Proceedings of the 19th international conference on World wide web* (pp. 591-600). ACM. 10.1145/1772690.1772751

Lampe, C., Ellison, N., & Steinfield, C. (2006, November). A Face (book) in the crowd: Social searching vs. social browsing. In *Proceedings of the 2006 20th anniversary conference on Computer supported cooperative work* (pp. 167-170). ACM. 10.1145/1180875.1180901

Lampe, C. A., Ellison, N., & Steinfield, C. (2007, April). A familiar face (book): profile elements as signals in an online social network. In *Proceedings of the SIGCHI conference on Human factors in computing systems* (pp. 435-444). ACM. 10.1145/1240624.1240695

Lange, P. G. (2007). Publicly private and privately public: Social networking on YouTube. *Journal of Computer-Mediated Communication, 13*(1), 361–380. doi:10.1111/j.1083-6101.2007.00400.x

Livingstone, S. (2008). Taking risky opportunities in youthful content creation: Teenagers' use of social networking sites for intimacy, privacy and self-expression. *New Media & Society, 10*(3), 393–411. doi:10.1177/1461444808089415

Ma, J., Qiao, Y., Hu, G., Huang, Y., Wang, M., Sangaiah, A. K., ... Wang, Y. (2017). Balancing User Profile and Social Network Structure for Anchor Link Inferring Across Multiple Online Social Networks. *IEEE Access: Practical Innovations, Open Solutions, 5*, 12031–12040. doi:10.1109/ACCESS.2017.2717921

Madejski, M., Johnson, M., & Bellovin, S. M. (2011). *The failure of online social network privacy settings.* Department of Computer Science, Columbia University, Tech. Rep. CUCS-010-11.

Mislove, A., Koppula, H. S., Gummadi, K. P., Druschel, P., & Bhattacharjee, B. (2008, August). Growth of the flickr social network. In *Proceedings of the first workshop on Online social networks* (pp. 25-30). ACM. 10.1145/1397735.1397742

Mislove, A., Marcon, M., Gummadi, K. P., Druschel, P., & Bhattacharjee, B. (2007, October). Measurement and analysis of online social networks. In *Proceedings of the 7th ACM SIGCOMM conference on Internet measurement* (pp. 29-42). ACM. 10.1145/1298306.1298311

Mislove, A., Viswanath, B., Gummadi, K. P., & Druschel, P. (2010, February). You are who you know: inferring user profiles in online social networks. In *Proceedings of the third ACM international conference on Web search and data mining* (pp. 251-260). ACM. 10.1145/1718487.1718519

Narendula, R., Papaioannou, T. G., & Aberer, K. (2011, August). My3: A highly-available P2P-based online social network. In *Peer-to-Peer Computing (P2P), 2011 IEEE International Conference on*(pp. 166-167). IEEE.

Nie, Y., Jia, Y., Li, S., Zhu, X., Li, A., & Zhou, B. (2016). Identifying users across social networks based on dynamic core interests. *Neurocomputing, 210*, 107–115. doi:10.1016/j.neucom.2015.10.147

Nosko, A., Wood, E., & Molema, S. (2010). All about me: Disclosure in online social networking profiles: The case of FACEBOOK. *Computers in Human Behavior, 26*(3), 406–418. doi:10.1016/j.chb.2009.11.012

Park, N., Kee, K. F., & Valenzuela, S. (2009). Being immersed in social networking environment: Facebook groups, uses and gratifications, and social outcomes. *Cyberpsychology & Behavior*, *12*(6), 729–733. doi:10.1089/cpb.2009.0003 PMID:19619037

Pempek, T. A., Yermolayeva, Y. A., & Calvert, S. L. (2009). College students' social networking experiences on Facebook. *Journal of Applied Developmental Psychology*, *30*(3), 227–238. doi:10.1016/j.appdev.2008.12.010

Raacke, J., & Bonds-Raacke, J. (2008). MySpace and Facebook: Applying the uses and gratifications theory to exploring friend-networking sites. *Cyberpsychology & Behavior*, *11*(2), 169–174. doi:10.1089/cpb.2007.0056 PMID:18422409

Rezaee, S., Lavesson, N., & Johnson, H. (2012). E-mail prioritization using online social network profile distance. *Computer Science & Applications*, *9*(1), 70–87.

Silfverberg, S., Liikkanen, L. A., & Lampinen, A. (2011, March). I'll press play, but I won't listen: profile work in a music-focused social network service. In *Proceedings of the ACM 2011 conference on Computer supported cooperative work* (pp. 207-216). ACM. 10.1145/1958824.1958855

Singh, R. R., & Tomar, D. S. (2009). *Approaches for user profile investigation in orkut social network*. arXiv preprint arXiv:0912.1008

Strater, K., & Lipford, H. R. (2008, September). Strategies and struggles with privacy in an online social networking community. In *Proceedings of the 22nd British HCI Group Annual Conference on People and Computers: Culture, Creativity, Interaction* (vol. 1, pp. 111-119). British Computer Society.

Strufe, T. (2010, April). Profile popularity in a business-oriented online social network. In *Proceedings of the 3rd workshop on social network systems* (p. 2). ACM. 10.1145/1852658.1852660

Subrahmanyam, K., Reich, S. M., Waechter, N., & Espinoza, G. (2008). Online and offline social networks: Use of social networking sites by emerging adults. *Journal of Applied Developmental Psychology*, *29*(6), 420–433. doi:10.1016/j.appdev.2008.07.003

Utz, S. (2010). Show me your friends and I will tell you what type of person you are: How one's profile, number of friends, and type of friends influence impression formation on social network sites. *Journal of Computer-Mediated Communication*, *15*(2), 314–335. doi:10.1111/j.1083-6101.2010.01522.x

Utz, S., & Beukeboom, C. J. (2011). The role of social network sites in romantic relationships: Effects on jealousy and relationship happiness. *Journal of Computer-Mediated Communication*, *16*(4), 511–527. doi:10.1111/j.1083-6101.2011.01552.x

Wischenbart, M., Mitsch, S., Kapsammer, E., Kusel, A., Pröll, B., Retschitzegger, W., ... Lechner, S. (2012, April). User profile integration made easy: model-driven extraction and transformation of social network schemas. In *Proceedings of the 21st International Conference on World Wide Web* (pp. 939-948). ACM. 10.1145/2187980.2188227

Zhang, H., Kan, M., Liu, Y., & Ma, S. (2014, November). Online social network profile linkage based on cost-sensitive feature acquisition. In *Chinese National Conference on Social Media Processing* (pp. 117-128). Springer Berlin Heidelberg. 10.1007/978-3-662-45558-6_11

Zhang, H., Kan, M. Y., Liu, Y., & Ma, S. (2014, December). Online social network profile linkage. In *Asia Information Retrieval Symposium* (pp. 197-208). Springer.

Zhou, X., Liang, X., Zhang, H., & Ma, Y. (2016). Cross-platform identification of anonymous identical users in multiple social media networks. *IEEE Transactions on Knowledge and Data Engineering*, *28*(2), 411–424. doi:10.1109/TKDE.2015.2485222

Chapter 6
Effects of Social Media Marketing Strategies on Consumers Behavior

Shamsher Singh
Banarsidas Chandiwala Institute of Professional Studies, India

Deepali Saluja
Banarsidas Chandiwala Institute of Professional Studies, India

ABSTRACT

In the information age, social media is growing rapidly and at a faster pace. Social media is playing an important role in the day-to-day life of individuals. Using social media has become the everyday routine. Many social media sites display different types of advertisements by which the decision-making process is generally getting affected. Social media is much more than just a medium of sharing information. The present study is an attempt to understand how social media affects the decision-making process of consumers and the impacts of various marketing strategies used by firms on social media. The study employs the survey method to collect primary data from 200 customers who have been regularly using social media. Factor analysis and ANOVA has been used to gain insights in the study. The selected respondents are assumed to represent the population in the urban areas of Delhi.

DOI: 10.4018/978-1-5225-5715-9.ch006

INTRODUCTION

What is the similarity between Ex US President Barack Obama, Indian super star Amitabh Bachchan, Indian Minister Narender Modi, present US President Donald Trump and Chinese seasoning and processed food maker Chings Secret? All these celebrity are using social networking for their brand promotion. Narender Modi has 40 million follower on face book followed by Donald Trump with 20 million, (Quartz India, 2017) Obama has a fan following of over a million, Bachchan of over three lakhs, Chings Secret has built over one lakh followers through social media. Today everyone is using some social networking site or the other. If they are a professional they use LinkedIn. If they have a flair for writing, they are either a regular blogger or use micro blogging sites like Twitter. If they are interested in connecting with their old friends or finding new ones they are likely to use Facebook, Orkut and many others. And there are several other social networking sites which cater to varied tastes, like Flickr for photography and YouTube for videos, music and movies. Invariably, more often than not, users on these sites are discussing a brand / product / service. These individuals are using social media to share views, and news about brands. Consequently, from the point of a brand promotion and management, social media becomes a significant tool.

Social media are media for social interaction, using highly accessible and scalable publishing techniques. Social media uses web-based technologies to turn communication into interactive dialogues. Kaplan and Heinlein (2010) define social media as "a group of Internet-based applications that build on the ideological and technological foundations of Web 2.0, which allows the creation and exchange of user-generated content". Social media is the medium to socialize. They use web-based technology to quickly disseminate knowledge and information to a huge number of users. They allow creation and exchange of user-generated content. Facebook, Twitter, Hi5, Orkut and other social networking sites are collectively referred to as social media. Social media represents low-cost tools that are used to combine technology and social interaction with the use of words. These tools are typically internet or mobile based like Twitter, Facebook, MySpace and YouTube.

Social Media, today, is among the 'best opportunities available' to a brand for connecting with prospective consumers. Social Media Marketing is the new mantra for several brands since early last year. Significantly different from conventional marketing strategies, Social Media Marketing (SMM) offers three distinct advantages.

- It provides a window to marketers to not only present products / services to customers but also to listen to customers' grievances and suggestions.
- It makes it easy for marketers to identify various peer groups or influencers among various groups, who in turn can become brand evangelist and help in organic growth of a brand.
- All this is done at nearly zero cost (as compared to conventional customer outreach programmes) as most of the social networking sites are free.

Social media marketing helps in generating exposure to businesses, increasing traffic /subscribers, building new business partnerships, rise in search engine rankings, generating qualified leads due to better lead generation efforts, selling more products and services and reduction in overall marketing expenses. The use of social media sites as part of a company's marketing strategy has increased significantly in the past couple years. As Swedowsky (2009) stated, businesses can not afford to ignore the benefits of using social media. In the past, consumers often just had the opinions of a few friends before making a significant purchase. The use of social media can increase number of opinions from just a few to hundreds or even thousands. Swedowsky reiterated that social media continues to abound for both businesses and the consumer.

Online access is no longer a luxury, it is a necessity. Businesses have also realized that consumers use social media because it is fun. They can easily share their ideas, photos, videos, likes and dislikes with each other. Businesses realize that importance of having increased interaction with consumers and retailers, and the use of social media gives them the opportunity to more efficiently meet the demand of their customers.

Many firms now use social media to enhance their marketing scheme. Firms also use social media for promotions and to survey groups for past purchases and interests. One has to be careful, however, when reading the reviews of any firm on a social media outlet. It is obvious that the use of social media to enhance marketing, is here to stay, so one must consider all possible avenues to positively use it to increase advertising and improve marketing. It is also obvious that there are benefits, drawbacks, and challenges associated with any social media strategy, and these must be addressed before a specific social media strategy is implemented. The purpose of social media should be to enhance a business branding and permit their biggest fans (i.e., super fans) to just talk about them. Businesses need to assist in facilitating the social media inputs and discussions.

The present article has made an attempt to find how the social media influences the consumer behaviour in general and in the urban area like Delhi in particular. The present study has used the survey method to collect the primary data from 200 respondents in Delhi who has been using the social media at least for last 6 months and are well versed with the uses of social media by the different organisations.

LITERATURE REVIEW

Social media includes online networks (e.g., Facebook, MySpace, and LinkedIn), wikis (e.g., Wikipedia), multimedia sharing sites (e.g., YouTube and Flickr), bookmarking sites (e.g., Del.icio.us and Digg), virtual worlds (e.g., Second Life), and rating sites (e.g. Yelp) (Edwards, 2011). The distinctive characteristic of social media is that it is a personalized user generated media. Users exercise greater control over its use and content generation (Dickey and Lewis, 2011).

Consumers are no more willing to listen what business organizations want them to listen rather they want business organizations to listen what they say. This attitudinal and behavioral transition in consumers is the impact of social media emergence and it is a big challenge for business firms to deal with it (Kietzmann et al., 2011). This situation signals that business firms should identify those factors of social media that affect the consumer attitude towards the product related information embedded in social media content. This may enable businesses to develop the affective social media promotional strategies. Social media has become the core of the marketing communication as some business gurus say that if business firms do not participate in social media they are not part of cyberspace anymore. Social media enables businesses to contact the end users directly and timely relatively at lower cost than traditional media (Kaplan and Haenlein, 2010).

With the rise in social networks, a new era of content creation has emerged, where individuals can easily share experiences and information with other users (Chen. 2011a) Social media provide opportunities for businesses to become more attractive universally (Chen 2011b). Today a large number of social media platforms have been developed that smooth the progress of sharing information and generation of content in an online context (Chen et al. 2011a). There are a number of social media that facilitate these activities, such as Wikipedia, Facebook, YouTube and Twitter. Individuals apply different social media tools, such as online forums and communities, recommendations, ratings and reviews, to interact with other users online. In fact, individuals are attracted online to exchange information and receive social support (Ridings & Gefen 2004).

Since the inception of social media, various studies have been conducted to examine its different aspects particularly those that drive the individuals to participate in social media. For instance, Daugherty et al. (2008) attempted to explore the factors motivating consumers to create social media content. Cheong and Morrison (2008) examined the consumers opinions about the recommendations and information implanted in user-generated content and producer generated content. Sun et al. (2009) studied the factors that support or inhibit users' knowledge sharing intentions in virtual. Imran and Zaheer (2012) in their study of social media theories argues that if more users of the social media produce the same views in the form of blogs, posts, scraps, reviews, comments etc., about a product it makes product related information credible. This postulation is in line with the assumption of social impact theory (Latane, 1981) which proposes as the number of people increases in a social group influence on target individual's attitude and behavior increases.

Social media users (e.g. members of Facebook) often share their product or brand related experiences and information in the form of posts, comments, and ratings. Individuals are influenced by the actions of others, entertained by their performance and sometimes persuaded by their arguments (Latane, 1981). Consumers rely on multiple sources to determine the credibility of information produced by online communities and bloggers. (Flanagin et al., 2011). (Cheung, Lee, and Rabjohn, 2008; Metzger, et al 2010).In addition, they also rely on the ratings of others, the number of posts, and the usefulness of the information presented (O'Reilly and Marx, 2011). User-generated information in the form of ratings and recommendations from others helps a consumer to assess the credibility of user generated contents (UGC). People join online social networks to search for technical specification of their desired products and brands, check out the new collections in different product categories, people read product related reviews or threads on social media to make the well-considered buying decisions (Muntinga, et al., 2011).

Social media explosion has changed the communication landscape around the globe (Edwards, 2011). It had affected the marketer and consumer relationship. Consumers are no more willing to listen what business organizations want them to listen rather they want business organizations to listen what they say (Kietzmann et al., 2011). Today young consumers believe more in the product related content or information created which other consumers generate on social networking sites, multi media sites, blogs, and so on than producer or company produced product related content, despite being personally unknown or unrelated to the user (Jonas, 2010). Nowak, et al (1990) suggested that a simple model of individual influence, operating in accordance with some general principles of social impact, if extended to reflect how individuals influence and are influenced by each other over time, can lead to plausible predictions of public opinion. Williams and Williams (1989) postulated that social impact varies depending on the underlying purpose of compliance.

Previous research has demonstrated that traditional media publicity can affect marketing outcomes (Agrawal and Kamakura 1995; Elberse 2007; Trusov et al., 2009), similarly online consumer-generated content such as online reviews can also affect sales (e.g.. Chevalier and Mayzlin 2006), and sometimes even negative publicity can have a positive marketing effect (Ahluwalia et al 2000; Berger, et al., 2010). Users perceive advertising differently depending on the social network, which suggests user motivations for online social networking may play a vital role in defining consumer's responses to social media marketing According to Chi (2011) social media is not only for advertising, but it can also be a tool for brands or services to connect with their consumers. Social media allows consumers and prospective consumers to communicate directly to a brand representative. Since most consumers are using the social media as tool to search and purchase items, brands or services use this advantage to advertise their products. The analysis of consumer behavior is central for marketing success, especially since most potential consumers are using the internet and different online socializing tools. The online audience is a booming market worldwide, however giving its globalized nature a level of segmentation is needed cross-culturally (Vinerean, et,.al, 2013)

Social media can build brand attitudes that affect buying behavior. The good image of brand or product can lead the consumer to make decision on their purchases. When consumer's friend on social media shares or recommends services or products on their social media, it affects brand attitude and influences their decision-making. Yet, advertising on social media, which is provided by commercial sources affect both consumer brand attitudes and purchasing intention (Yang, 2012). From that information, it helps marketers plan their marketing strategies. Many marketers use social media for marketing campaigns. It is the easy way to communicate with consumers; also it is inexpensive to advertise their brands or services.

In the marketing literature, scholars have specifically emphasized the effects of online word of mouth through social media activities (Kimmel and Kitchen 2014). Organizations have realized social media to be a new set of business processes and operations that may help firms increase sales (Dewan and Ramaprasad 2014). The study of Abdul Razak,and Latip.(2016) revealed that there are a few factors that influences the usage of social media marketing by SMEs. These factors were identified as usefulness, ease of use and enjoyment.

Areeba, et,.al, (2017) found that marketers need to consider the strategic role of consumer engagement in arousing purchase intention. They also need to strategically enhance their social media marketing communication so that the maximum benefits of engaging customers can be reaped. Marketers should consider offering some additional support to consumers through their social media page design, for example, allow space for consumers to share their related experiences and opinions of using certain products or services with other consumers. This kind of information can help other consumers to decide on what they want to purchase.

Jack Ma, the founder of Chinese e-commerce firm, Alibaba, said that failure to utilise the social network platform as a media to interact with consumers and other business associates or prospects and potentials may lead them to be excluded from the industry, thereby, losing their position in the market (Barhemmati & Ahmad, 2015). Consumer engagement on social networking sites is largely supported by an emotional attachment which is directed to enhance their purchase behavior. Such an engagement would lead to increased consumer loyalty who can then promote the brand and its products to other consumers in the virtual world (Asperen, Rooij, & Dijkmans, 2017)

Dehghani and Tumer (2015) find that Facebook advertising can significantly affect the brand image and brand equity by offering greater interactivity, personalisation and feedback. This process can in turn, affect consumer purchase intentions. Pjero and Kercini (2015) in their study focusing on social media and its influence on consumer behaviour observe that information about products and services offered in the virtual world can positively impact the purchase intentions of consumers. A consumer may be influenced by eWOM (electronic word-of-mouth) by other users.

In today's borderless world, people throughout the globe are using social media in real time for various purposes of communication, with majority spending almost a quarter of their daily time, surfing social networks Companies offering services and products are determined to get the attention of social network consumers thus, these companies are redesigning their marketing strategies and policies. Among these is the strategy of integrating social media into their marketing scheme, one aspect of change that is hoped to project their products, services and brands to the outside world(Forbes, 2017).

The advertising quality on the social networking website is an important factor from user point of view. Social media has strong influence and impact on decision making of customers, however many a time the information available on social media website does not make user comfortable while making decision and they have to search for other sources of information before making buying decision. The organisation likely to use social media as part of their marketing strategies must ensure that sufficient information is provided to the uses so that they are comfortable in making purchase decision (Singh,2016)

There has been little research to explore the implications for the sales force on social media and the communication technology usage patterns associated with it. It would seem that this is a vital gap for sales force research and practice. In particular, social media, such as Facebook, LinkedIn, and now Google+, and communication technology, such as mobile Internet and the smart phone, appear to be more than simple extensions of traditional technology such as static phones, desktops, and even laptops. Today, salespeople often do not have a choice as to whether or not they are contactable. The rise of global business exacerbates such a situation, meaning salespeople might feel that they are on call 24 hours a day and are expected to respond to communications immediately whatever time they come through from anywhere in the world. (Marshall,et al., 2012).

Customer engagement has become a "strategic imperative" for businesses since "engaged customers play a key role in viral marketing activity by providing referrals and/or recommendations for specific products, services, and/or brands to others" (Brodie et al. 2011). Customer-oriented salespeople have often sought new ways to interact and engage with customers to co-create value (Jones, et al., 2003). Bowden (2009) suggests that customer engagement involves a calculative commitment, which forms the basis for consequent buying behavior; leads to an emotional commitment, which forms the basis for buyer loyalty and increases buyers' involvement and trust.

Customer engagement is often considered to be highly correlated with trust and commitment in buyer–seller relationships. Push strategies that utilize large professional networks to disseminate referrals are likely to be more effective when customers are highly engaged. Customer engagement can be increased by providing opportunities to seek, give, and pass opinions (Chu and Kim 2011) about products and services in online communities, wikis, and creative works–sharing sites (e.g., YouTube, Flickr). Such co-created value represents a dynamic and swelling fountain of knowledge for existing and future customers, which can be used to increase buyer involvement and trust.

OBJECTIVES AND HYPOTHESES

This study seeks to find out the impact of social media marketing strategies on consumer behaviour . In pursuance of the above objectives, the following hypotheses were formulated for testing:

$H_{01:}$ There are no significant variations in the customer response for social media marketing strategies based on respondents' gender.

H_{02}: There are no significant variations in the customer response for social media marketing strategies based on respondents' age group.

H_{03}: There are no significant variations in the customer response for social media marketing strategies based on respondents' education level.

H_{04}: There are no significant variations in the customer response for social media marketing strategies based on respondents' occupation.

H_{05}: There are no significant variations in the customer response for social media marketing strategies based on respondents' income.

H_{06}: The advertising quality on the social networking website in not an important factor for decision making by consumers.

H_{07}: Social Media does not have any impact on professional or personal life of users.

H_{08}: The content of social media does not have any impact on decision making by consumers.

H_{09}: Social media does not have any influence on decision making process of consumers related to any field.

H_{010}: The information on social media website does not make user comfortable while making decision.

H_{011}: Brand/ Organisation present on social media website does not have any influences on decision making process by consumers.

METHODOLOGY

The study has employed both primary and secondary data. The secondary data has been used to gain in the in depth knowledge of social media networking sites, where as primary data has been collected from the social media user to find their responses on the various aspects of social media marketing.

Based on the literature review on different aspects of social media a well structured questionnaire was developed. Prior to the final survey, the questionnaire was pre tested using a sample of respondents similar in nature to the final sample. The goal of pilot survey was to ensure readability and logical arrangements of questions. The final questionnaire was administered to 200 customers who were using the one or other social media sites by means of face-to-face interviews.

Limitations of the Study

The study has been carried out in the National Capital Region, which is metropolitan city area where the education level and income level of population is very high as compare to rest of India. Also there is high penetration of internet user especially in the young generation which uses the social media networking site for their personal as well professional uses. The finding of the study cannot be generalised for the whole country, hence the finding of the research can be generalised only for urban areas only.

Research and Statistical Tools Employed

The research and statistical tools employed in this study are, ANOVA, T test and Factor Analysis . SPSS 16 was used to perform statistical analysis. The reliability of the data was carried out by using Cronbach's Alpha Value. T Test and ANOVA was employed to find the significant factor which will determine the overall customer satisfaction. The third major analysis carried out was a factor analysis to examine the underlying or latent dimensions within variables of overall satisfaction (Hair, Anderson, Tatham and Black, 1998). Both Bartlett's test of spherecity and measure of sampling adequacy (MSA) were also carried out to ensure that the requirements of factor analysis were met.

DATA ANALYSIS AND INTERPRETATION

The analysis of this data was divided into following sections:

1. **Demographic profile of Respondents:** Table 1.
2. **Reliability and Validity:** Table 2.
3. **Factor Analysis:** Table 3,4&5.
4. **ANOVA:** Table 6 to 10.
5. **T- Test Computation:** Table 11.

Table 1 indicates the respondent's profile. There are 57% Male and 43% female. Students are largest respondents (4%) followed by Govt. Service (27%) and private service (24%) . Most of the respondents are either graduate (36%) or post graduate (40.%). There are 48% respondents in the 15-25 years age category, 30% in the 25-40 years category and only 22% respondents are in above 40 years. The profile of respondents indicates they are young, educated and decently employed. They belong to the new generation who are using the social networking sites for different purposes such as to connect with their friends, relatives & peers, for networking to make a opinion or to express their view point .

RELIABILITY AND VALIDITY

Table 2 shows the result of reliability analysis- Cronbach's Alpha Value. This test measured the consistency between the survey scales. The Cronbach's Alpha score of 1.0 indicate 100 percent reliability. Cronbach's Alpha scores were all greater than

Table 1. Demographic profile of Respondents

Variable	Characteristics	Frequency	Percentage
Gender	Male	114	57
	Female	86	43
Occupation	Government service	54	27
	Private service	48	24
	Student	98	49
Education	Post Graduation	80	40
	Graduation	72	36
	10+2	40	20
	Others	8	4.0
Age group	15-25 yrs	96	48
	25-40 yrs	60	30
	40 yrs & above	44	22
Annual Income	"Less than Rs 100000"	58	29
	Rs.1000001to 300000	62	31
	Rs 300001to 500000	50	25
	Rs 500001and above	30	15
	Total	**200**	**100**

Table 2. Reliability Statistics

Cronbach's Alpha	Cronbach's Alpha Based on Standardized Items	N of Items
.775	.723	19

the Nunnaly's (1978) generally accepted score of 0.7. In this case, the score was 0.775 for the service quality provided by the banks.

Factor Analysis

Overall, the set of data meets the fundamental requirements of factor analysis satisfactorily (Hair et al, 1998). The Kaiser-Meyer-Olkin Measure of sampling adequacy is 0.522 which is above the acceptable level of 0.5 In analyzing the data given, the 14 response items were subjected to a factor analysis using the principal component method. Using the criteria of an Eigen value greater than one, six clear factors emerged accounting for 60.20% of the total variance. As in common practice, a Varimax rotation with Kaiser Normalization was performed to achieve a simpler and theoretically more meaningful factor solution. The Cronbach alphas score for all the factors was 0.775 (Table 2). There are six clear factor emerges from factor loadings as highlighted in Table 10. These six factors represent different elements of social networking site that form the underlying factors from the original 14 scale response items given.

Referring to the Table 10, first factor represents elements of social networking sites directly related to availability of informations, therefore it is labeled as "Information factor". These elements are "social media website as source of informations, quality of advertisement on websites and field of decision making". Second factor is directly

Table 3. KMO and Bartlett's Test

Kaiser-Meyer-Olkin Measure of Sampling Adequacy.		.522
Bartlett's Test of Sphericity	Approx. Chi-Square	102.762
	df	91
	Sig.	.188

Table 4. Total variance explained

Component	Initial Eigenvalues			Extraction Sums of Squared Loadings			Rotation Sums of Squared Loadings		
	Total	% of Variance	Cumulative %	Total	% of Variance	Cumulative %	Total	% of Variance	Cumulative %
1	1.783	12.733	12.733	1.783	12.733	12.733	1.559	11.139	11.139
2	1.591	11.364	24.097	1.591	11.364	24.097	1.424	10.168	21.307
3	1.406	10.043	34.140	1.406	10.043	34.140	1.373	9.806	31.113
4	1.337	9.550	43.691	1.337	9.550	43.691	1.362	9.730	40.843
5	1.183	8.448	52.139	1.183	8.448	52.139	1.357	9.695	50.538
6	1.129	8.067	60.206	1.129	8.067	60.206	1.354	9.669	60.206
7	.930	6.642	66.849						
8	.804	5.741	72.590						
9	.780	5.570	78.160						
10	.728	5.201	83.361						
11	.696	4.971	88.331						
12	.619	4.423	92.755						
13	.545	3.893	96.648						
14	.469	3.352	100.000						
Extraction Method: Principal Component Analysis.									

Table 5. Rotated Component Matrix[a]

Characteristics	Component					
	1	2	3	4	5	6
Social Media Website	-.026	-.107	.196	-.026	.287	**.690**
Favourite Social Media Site	.114	.165	**.604**	-.381	.135	.043
Time Spent on site	.277	.292	-.047	.015	**.671**	.131
Preferred Time Slot of using social media website	.048	-.773	.196	-.023	.001	.139
Specific Reason for using social media website	-.089	**.744**	.325	.054	.052	.017
Social media website as Source of Informations	**.495**	-.253	-.014	-.140	.239	-.339
Type of Advertisement viewed	-.043	-.045	-.152	.032	-.067	**.610**
Quality of Advertisement on websites	**.634**	-.150	-.170	.297	-.041	.152
Impact on the Professional Life	.408	-.061	.320	**.609**	-.095	.202
Impact on Decision Making	.102	.087	-.028	**.633**	-.341	.354
Influence of Social media website on Decision Making	.263	.059	-.612	**.364**	.091	.169
Field of Decision Making	**.712**	.058	.153	-.035	-.054	-.091
Comfort of making Purchasing Decision	-.187	-.117	.022	.105	**.777**	.008
Impact of Brand or Organisation	.147	.166	.029	**.733**	.006	.146
Extraction Method: Principal Component Analysis. Rotation Method: Varimax with Kaiser Normalization. [a] Rotation converged in 17 iterations						

related to specific reason for using social media website; it is therefore labeled as "User factor". Third factor is directly related to favourite social media site and therefore named as "Preference factor". Fourth factor represent the impact of social media website therefore it is labeled as "Impact factor" These elements are "impact on the professional life, impact on decision making, influence of social media website on decision making and field of decision making". Fifth factor is related to time spent on website and, comfort of making purchasing decision therefore it is named as "Decision factor". Sixth factor is related to type of social media website and type of advertisement viewed on these website and therefore it is labeled as "Advertisement factor".

Table 6. Computation of ANOVA on the basis of gender

		Sum of Squares	df	Mean Square	F	Sig.
Social Media Website	Between Groups	8.876	1	8.876	2.901	.092
	Within Groups	299.874	98	3.060		
	Total	308.750	99			
Favourite Social Media Site	Between Groups	.160	1	.160	.044	.835
	Within Groups	357.880	98	3.652		
	Total	358.040	99			
Time Spent on site	Between Groups	.253	1	.253	.411	.523
	Within Groups	60.257	98	.615		
	Total	60.510	99			
Preferred Time Slot of using social media website	Between Groups	3.682	1	3.682	2.518	.116
	Within Groups	143.318	98	1.462		
	Total	147.000	99			
Specific Reason for using social media website	Between Groups	3.815	1	3.815	.652	.421
	Within Groups	573.575	98	5.853		
	Total	577.390	99			
Social media website as Source of Informations	Between Groups	.913	1	.913	.405	.526
	Within Groups	220.877	98	2.254		
	Total	221.790	99			
Type of Advertisement viewed	Between Groups	1.948	1	1.948	.895	.347
	Within Groups	213.362	98	2.177		
	Total	215.310	99			
Quality of Advertisement on websites	Between Groups	2.406	1	2.406	1.146	.287
	Within Groups	205.834	98	2.100		
	Total	208.240	99			
Impact on the Professional Life	Between Groups	.235	1	.235	.112	.739
	Within Groups	205.765	98	2.100		
	Total	206.000	99			

continued on following page

Table 6. Continued

		Sum of Squares	df	Mean Square	F	Sig.
Impact on Decision Making	Between Groups	.940	1	.940	.526	.470
	Within Groups	175.060	98	1.786		
	Total	176.000	99			
Influence of Social media website Decision Making	Between Groups	1.810	1	1.810	.897	.346
	Within Groups	197.750	98	2.018		
	Total	199.560	99			
Field of Decision Making	Between Groups	.001	1	.001	.000	.983
	Within Groups	320.109	98	3.266		
	Total	320.110	99			
Comfort of making Purchasing Decision	Between Groups	.901	1	.901	.513	.475
	Within Groups	172.099	98	1.756		
	Total	173.000	99			
Impact of Brand or Organisation	Between Groups	2.150	1	2.150	1.094	.298
	Within Groups	192.610	98	1.965		
	Total	194.760	99			

Table 7. Computation of ANOVA on the basis of age group

		Sum of Squares	df	Mean Square	F	Sig.
Social Media Website	Between Groups	17.257	5	3.451	1.113	.359
	Within Groups	291.493	94	3.101		
	Total	308.750	99			
Favourite Social Media Site	Between Groups	23.313	5	4.663	1.309	.267
	Within Groups	334.727	94	3.561		
	Total	358.040	99			
Time Spent on site	Between Groups	6.459	5	1.292	2.247	.056
	Within Groups	54.051	94	.575		
	Total	60.510	99			

continued on following page

Table 7. Continued

		Sum of Squares	df	Mean Square	F	Sig.
Preferred Time Slot of using social media website	Between Groups	3.214	5	.643	.420	.834
	Within Groups	143.786	94	1.530		
	Total	147.000	99			
Specific Reason for using social media website	Between Groups	30.635	5	6.127	1.053	.391
	Within Groups	546.755	94	5.817		
	Total	577.390	99			
Social media website as Source of Informations	Between Groups	12.025	5	2.405	1.078	.378
	Within Groups	209.765	94	2.232		
	Total	221.790	99			
Type of Advertisement viewed	Between Groups	22.736	5	4.547	2.220	.059
	Within Groups	192.574	94	2.049		
	Total	215.310	99			
Quality of Advertisement on websites	Between Groups	5.556	5	1.111	.515	.764
	Within Groups	202.684	94	2.156		
	Total	208.240	99			
Impact on the Professional Life	Between Groups	21.238	5	4.248	2.161	.065
	Within Groups	184.762	94	1.966		
	Total	206.000	99			
Impact on Decision Making	Between Groups	6.917	5	1.383	.769	.574
	Within Groups	169.083	94	1.799		
	Total	176.000	99			
Influence of Social media website Decision Making	Between Groups	10.552	5	2.110	1.050	.393
	Within Groups	189.008	94	2.011		
	Total	199.560	99			
Field of Decision Making	Between Groups	6.436	5	1.287	.386	.857
	Within Groups	313.674	94	3.337		
	Total	320.110	99			
Comfort of making Purchasing Decision	Between Groups	5.568	5	1.114	.625	.681
	Within Groups	167.432	94	1.781		
	Total	173.000	99			
Impact of Brand or Organisation	Between Groups	12.209	5	2.442	1.257	.289
	Within Groups	182.551	94	1.942		
	Total	194.760	99			

Table 8. Computation of ANOVA on the basis of education

		Sum of Squares	df	Mean Square	F	Sig.
Social Media Website	Between Groups	15.994	3	5.331	1.748	.162
	Within Groups	292.756	96	3.050		
	Total	308.750	99			
Favourite Social Media Site	Between Groups	5.276	3	1.759	.479	.698
	Within Groups	352.764	96	3.675		
	Total	358.040	99			
Time Spent on site	Between Groups	2.030	3	.677	1.111	.349
	Within Groups	58.480	96	.609		
	Total	60.510	99			
Preferred Time Slot of using social media website	Between Groups	5.460	3	1.820	1.234	.302
	Within Groups	141.540	96	1.474		
	Total	147.000	99			
Specific Reason for using social media website	Between Groups	44.914	3	14.971	2.699	.050
	Within Groups	532.476	96	5.547		
	Total	577.390	99			
Social media website as Source of Informations	Between Groups	4.167	3	1.389	.613	.608
	Within Groups	217.623	96	2.267		
	Total	221.790	99			
Type of Advertisement viewed	Between Groups	3.686	3	1.229	.557	.644
	Within Groups	211.624	96	2.204		
	Total	215.310	99			
Quality of Advertisement on websites	Between Groups	3.272	3	1.091	.511	.676
	Within Groups	204.968	96	2.135		
	Total	208.240	99			
Impact on the Professional Life	Between Groups	4.598	3	1.533	.731	.536
	Within Groups	201.402	96	2.098		
	Total	206.000	99			

continued on following page

Table 8. Continued

		Sum of Squares	df	Mean Square	F	Sig.
Impact on Decision Making	Between Groups	4.068	3	1.356	.757	.521
	Within Groups	171.932	96	1.791		
	Total	176.000	99			
Influence of Social media website Decision Making	Between Groups	5.167	3	1.722	.851	.470
	Within Groups	194.393	96	2.025		
	Total	199.560	99			
Field of Decision Making	Between Groups	7.839	3	2.613	.803	.495
	Within Groups	312.271	96	3.253		
	Total	320.110	99			
Comfort of making Purchasing Decision	Between Groups	2.214	3	.738	.415	.743
	Within Groups	170.786	96	1.779		
	Total	173.000	99			
Impact of Brand or Organisation	Between Groups	5.560	3	1.853	.940	.424
	Within Groups	189.200	96	1.971		
	Total	194.760	99			

Table 9. Computation of ANOVA on the basis of occupation

		Sum of Squares	df	Mean Square	F	Sig.
Social Media Website	Between Groups	5.336	2	2.668	.853	.429
	Within Groups	303.414	97	3.128		
	Total	308.750	99			
Favourite Social Media Site	Between Groups	.969	2	.485	.132	.877
	Within Groups	357.071	97	3.681		
	Total	358.040	99			
Time Spent on site	Between Groups	1.429	2	.715	1.173	.314
	Within Groups	59.081	97	.609		
	Total	60.510	99			
Preferred Time Slot of using social media website	Between Groups	8.073	2	4.037	2.818	.065
	Within Groups	138.927	97	1.432		
	Total	147.000	99			

continued on following page

Table 9. Continued

		Sum of Squares	df	Mean Square	F	Sig.
Specific Reason for using social media website	Between Groups	19.615	2	9.808	1.706	.187
	Within Groups	557.775	97	5.750		
	Total	577.390	99			
Social media website as Source of Informations	Between Groups	8.528	2	4.264	1.939	.149
	Within Groups	213.262	97	2.199		
	Total	221.790	99			
Type of Advertisement viewed	Between Groups	2.504	2	1.252	.571	.567
	Within Groups	212.806	97	2.194		
	Total	215.310	99			
Quality of Advertisement on websites	Between Groups	7.713	2	3.856	1.865	.160
	Within Groups	200.527	97	2.067		
	Total	208.240	99			
Impact on the Professional Life	Between Groups	3.088	2	1.544	.738	.481
	Within Groups	202.912	97	2.092		
	Total	206.000	99			
Impact on Decision Making	Between Groups	4.006	2	2.003	1.130	.327
	Within Groups	171.994	97	1.773		
	Total	176.000	99			
Influence of Social media website Decision Making	Between Groups	1.800	2	.900	.441	.644
	Within Groups	197.760	97	2.039		
	Total	199.560	99			
Field of Decision Making	Between Groups	4.726	2	2.363	.727	.486
	Within Groups	315.384	97	3.251		
	Total	320.110	99			
Comfort of making Purchasing Decision	Between Groups	4.306	2	2.153	1.238	.294
	Within Groups	168.694	97	1.739		
	Total	173.000	99			
Impact of Brand or Organisation	Between Groups	1.788	2	.894	.449	.639
	Within Groups	192.972	97	1.989		
	Total	194.760	99			

Table 10. Computation of ANOVA on the basis of annual income

		Sum of Squares	df	Mean Square	F	Sig.
Social Media Website	Between Groups	3.749	3	1.250	.393	.758
	Within Groups	305.001	96	3.177		
	Total	308.750	99			
Favourite Social Media Site	Between Groups	12.340	3	4.113	1.142	.336
	Within Groups	345.700	96	3.601		
	Total	358.040	99			
Time Spent on site	Between Groups	3.713	3	1.238	2.092	.106
	Within Groups	56.797	96	.592		
	Total	60.510	99			
Preferred Time Slot of using social media website	Between Groups	4.853	3	1.618	1.093	.356
	Within Groups	142.147	96	1.481		
	Total	147.000	99			
Specific Reason for using social media website	Between Groups	14.850	3	4.950	.845	.473
	Within Groups	562.540	96	5.860		
	Total	577.390	99			
Social media website as Source of Informations	Between Groups	13.615	3	4.538	2.093	.106
	Within Groups	208.175	96	2.168		
	Total	221.790	99			
Type of Advertisement viewed	Between Groups	2.109	3	.703	.316	.813
	Within Groups	213.201	96	2.221		
	Total	215.310	99			
Quality of Advertisement on websites	Between Groups	4.058	3	1.353	.636	.594
	Within Groups	204.182	96	2.127		
	Total	208.240	99			
Impact on the Professional Life	Between Groups	6.970	3	2.323	1.121	.345
	Within Groups	199.030	96	2.073		
	Total	206.000	99			

continued on following page

Table 10. Continued

		Sum of Squares	df	Mean Square	F	Sig.
Impact on Decision Making	Between Groups	4.257	3	1.419	.793	.501
	Within Groups	171.743	96	1.789		
	Total	176.000	99			
Influence of Social media website Decision Making	Between Groups	3.933	3	1.311	.643	.589
	Within Groups	195.627	96	2.038		
	Total	199.560	99			
Field of Decision Making	Between Groups	1.564	3	.521	.157	.925
	Within Groups	318.546	96	3.318		
	Total	320.110	99			
Comfort of making Purchasing Decision	Between Groups	10.579	3	3.526	2.084	.107
	Within Groups	162.421	96	1.692		
	Total	173.000	99			
Impact of Brand or Organisation	Between Groups	5.443	3	1.814	.920	.434
	Within Groups	189.317	96	1.972		
	Total	194.760	99			

RESULTS AND IMPLICATIONS

ANOVA has been employed to find the significant factor of social media website based on the demographic factor and to test the first five hypotheses. The significance level of 0.05 has been used as cut off for either acceptance or rejection the hypothesis. The analysis of variance based on gender indicates that the significance level is higher than 0.05 hence we accept the $H_{01:}$ and conclude that there is no significant variation in the customer response for social media marketing strategies based on respondents' gender.

The analysis of variance based on respondents' age group indicate that significance level is either less or equal to 0.05 in the attribute such as "time spent on site, type of advertisement viewed and impact on the professional life, hence we reject H_{02} and conclude that there are significant difference in the customer response for social media marketing strategies based on respondents' age group. The marketing managers may consider this important aspects while designing their marketing strategies for social media website.

The analysis of variance based on respondents' education level shows that the significance level is equal to 0.05 in the attribute s "specific reason for using social media website", hence we reject $\mathbf{H_{03}}$ and conclude that there are significant difference in the customer response for social media marketing strategies based on respondents' education level. Respondents uses the social media website for different purposes such as making purchase decision, friendship, professional purposes etc.

The analysis of variance based on respondents' occupation shows that the significance level is more than 0.05 level for all attributes hence we accept the $\mathbf{H_{04}}$ and conclude that there are no significant difference in the customer response for social media marketing strategies site based on respondents' occupation . The analysis of variance based on respondents' income shows that the significance level is more than 0.05 level for all attributes hence we accept the $\mathbf{H_{05}}$ and conclude that are no significant variations in the customer response for social media marketing strategies based on respondents' income. This further strengthens the arguments that the marketer can have similar Marketing strategies for all income groups.

In order to test the remaining hypothesis the one sample T- test was carried out. This "significance value" is the P-value. It is the probability of a test score indicating the higher value or more [in the direction of the alternative hypothesis] if the null hypothesis is true. The smaller the number, the rarer our test score is under H_0 and the more likely that the null hypothesis isn't true. Using the .05 significance level as our cutoff, we find the P-value of .000 to be in our rejection region. Based on the significance value (.000) we reject null hypothesis H06 and conclude that the advertising quality on the social networking website is an important factor from user point of view. Similarly based on the significance value (.000) we reject null

Table 11. T- Test Computation

Statement	Test Value = 0					
	t	df	Sig. Value (2-tailed)	Mean Difference	95% Confidence Interval of the Difference	
					Lower	Upper
Interesting Advertisement Quality	15.445	99	.000	2.24000	1.9522	2.5278
Impact on Professional / Personal Life	15.251	99	.000	2.20000	1.9138	2.4862
Influence on Decision Making	16.763	99	.000	2.38000	2.0983	2.6617
Impact on Decision Making	21.938	99	.000	2.90000	2.6377	3.1623
Comfort in Purchasing Decision	19.500	99	.000	2.60000	2.3354	2.8646
Brand/ Organisation Influences on Decision making	18.519	99	.000	3.33000	2.9732	3.6868

hypothesis H07 and conclude that Social Media does have impact on professional or personal life of users. Similarly based on the significance value (.000) we reject the entire remaining null hypothesis. This indicate that he content of social media does have impact on decision making by consumers and influences their decision making. It seems that many a time, information available on social media website is not sufficient enough to make user comfortable while making decision but it does have influence in their decision making also the Brand/ Organisation present on social media website does influences their decision making .

FUTURE DIRECTION AND MANAGERIAL IMPLICATIONS

The study is descriptive in nature and provides the useful information regarding customer perspective of social media sites and how it is influencing their decision making. This study has brought an important finding that irrespective of gender, occupation and income of the customer, the marketing professional may design their marketing strategies which should have focus on broader segmentation or mass segmentation . The organisation must have their presence on the social networking sites in order to optimize their marketing promotion. The marketing only on the social media is not an ideal strategy; however it can be complementary to the traditional promotional media. Managers while deciding the promotional mix, they should include the promotion on social media websites as well. Future research can be carried out based on the findings of this study to establish its effect on customer decision making and how it affects the satisfaction. This research did not study the association between customer satisfaction and retention of customers. Additional research may well explore the relationship between these two constructs.

CONCLUSION

The empirical evidence based on ANOVA indicates that there are not significant variance in customer responses based on the gender, occupation and income. This indicate that the customer irrespective of their gender, occupation and income view the social media networking site in similar way. However there are significant variance on the bases of age and education of the respondents. This study bring out the significant characteristic of social media " such as time spent on site, type of advertisement viewed, specific reason for using social media website and impact on the professional life". The organisation should take in to account while designing their marketing strategies. The result of T- test indicate that advertising quality on the social networking website is an important factor from user point of view. Social media has strong influence and impact on decision making. The organisation likely to use social media as part of their marketing strategies may ensure that sufficient information is provided to the uses so that they are comfortable in making purchase decision. It was found that brand/ organisation which are present on social media website does influences purchase decision making of customers, which suggest that organisation should carefully choose the contents of the advertisement on social media. Factor analysis have brought six factor representing different elements of social networking sites These factors are "information factor, user factor, preference factor, impact factor, decision factor and advertisement factor". This brings out the important factor regarding how the manager should plan their marketing strategy depending upon the focus of the strategy.

REFERENCES

Abdul Razak, S.B., & Latip, N.A.B. (2016)."Factors That Influence The Usage of Social Media In Marketing. *Journal of Research in Business and Management*, *4*(2), 1-7.

Agrawal, J., & Kamakura, W. A. (1995). The Electronic Worth of Celebrity Endorsers: An Event Study Analysis. *Journal of Marketing*, *59*(July), 56–62. doi:10.2307/1252119

Ahluwalia, R., Robert, B., & Unnava, H. (2000). Consumer Response to Negative Publicity: The Moderating Role of Commitment. *JMR, Journal of Marketing Research*, *37*(2), 203–214. doi:10.1509/jmkr.37.2.203.18734

Asperen, M., Rooij, P., & Dijkmans, C. (2017). Engagement-based loyalty: The effects of social media engagement on customer loyalty in the travel industry. *International Journal of Hospitality & Tourism Administration*, *17*(4), 1–17.

Barhemmati, N., & Ahmad, A. (2015). Effects of social network marketing (SNM) on consumer purchase behavior through customer engagement. *Journal of Advanced Management Science*, *3*(4), 307–311. doi:10.12720/joams.3.4.307-311

Berger, J., Sorensen, A. T., & Rasmussen, S. J. (2010). Positive Effects of Negative Publicity: When Negative Reviews Increase Sales. *Marketing Science*, *29*(5), 815–827. doi:10.1287/mksc.1090.0557

Bowden, J. L.-H. (2009). The Process of Customer Engagement: A Conceptual Framework. *Journal of Marketing Theory and Practice*, *17*(1), 63–74. doi:10.2753/MTP1069-6679170105

Brodie, R. J., Hollebeek, L. D., Jurić, B., & Ilić, A. (2011). Customer Engagement: Conceptual domain, fundamental preposition and implication of research. *Journal of Service Research, 14*(3), 252–271. doi:10.1177/1094670511411703

Chen, J., Xu, H., & Whinston, A. B. (2011a). Moderated online communities and quality of user-generated content. *Journal of Management Information Systems, 28*(2), 237–268. doi:10.2753/MIS0742-1222280209

Chen, Y., Fay, S., & Wang, Q. (2011b). The role of marketing in social media: How online consumer reviews evolve. *Journal of Interactive Marketing, 25*(2), 85–94. doi:10.1016/j.intmar.2011.01.003

Cheong, H. J., & Morrison, M. A. (2008). Consumers' reliance on product information and recommendations found in UGC". *Journal of Interactive Advertising, 8*(2), 38–49. doi:10.1080/15252019.2008.10722141

Cheung, C. M. K., Lee, M. K. O., & Rabjohn, N. (2008). The impact of electronic word of-mouth: The adoption of online opinions in online customer communities. *Internet Research, 18*(3), 229–247. doi:10.1108/10662240810883290

Chevalier, J. A., & Mayzlin, D. (2006). The Effect of Word of Mouth on Sales: Online Book Reviews. *JMR, Journal of Marketing Research, 43*(August), 345–354. doi:10.1509/jmkr.43.3.345

Chi, H.-H. (2011). Interactive Digital Advertising VS. Virtual Brand Community: Exploratory Study of User Motivation and Social Media Marketing Responses in Taiwan. *Journal of Interactive Advertising, 12*(1), 44–61. doi:10.1080/15252019.2011.10722190

Chu, S. C., & Kim, Y. (2011). Determinants of Consumer engagement in electronic word of mouth (eWOM) in social networking sites". *International Journal of Advertising, 30*(1), 47–75. doi:10.2501/IJA-30-1-047-075

Daugherty, T., Eastin, M. S., & Bright, L. (2008). Exploring consumer motivations for creating user-generated content. *Journal of Interactive Advertising, 8*(2), 16–25. doi:10.1080/15252019.2008.10722139

Dehghani, M., & Tumer, M. (2015). A research on effectiveness of Facebook advertising on enhancing purchase intention of consumers. *Computers in Human Behavior, 49*(1), 597–600. doi:10.1016/j.chb.2015.03.051

Dewan, S., & Ramaprasad, J. (2014). Social Media, Traditional Media, and Music Sales. *Management Information Systems Quarterly, 38*(1), 101–121. doi:10.25300/MISQ/2014/38.1.05

Dickey, I. J., & Lewis, W. F. (2011). *An Overview of Digital Media and Advertising*. Information Science Reference.

Edwards, S. M. (2011). A social media mindset. *Journal of Interactive Advertising*, *12*(1), 1–3. doi:10.1080/15252019.2011.10722186

Elberse, A. (2007). The Power of Stars: Do Star Actors Drive the Success of Movies? *Journal of Marketing*, *71*(October), 102–120. doi:10.1509/jmkg.71.4.102

Flanagin, A. J., Metzger, M. J., Pure, R., & Markov, A. (2011). User-generated ratings and the evaluation of credibility and product quality in ecommerce transactions. In *Proceedings of the 44th Hawaii International Conference of System Sciences*. ACM Digital Library. 10.1109/HICSS.2011.474

Forbes. (2017). *4Tips To Help Your Business Flourish On Social Media*. Retrieved from https://www.forbes.com/sites/jpmorganchase/2017/03/20/4-tips-to-help-your-business-flourish-on-socialmedia/#605da1987dd2

Hair, J., Anderson, R., Tatham, & Black, W. (1998). Multivariate Data Analysis (5th ed.). PHI.

India, Q. (2017). *It's official, Narendra Modi is the most followed world leader on Facebook*. Accessed on 22 -10- 2017, retrieved from https://qz.com/917170/its-official-narendra-modi-is-the-most-followed-world-leader-on-facebook/

Jonas, J. R. O. (2010). Source credibility of company-produced and user generated content on the internet: An exploratory study on the Filipino youth. *Philippine Management Review*, *17*, 121–132.

Jones, E., Busch, P., & Dacin, P. (2003). Firm Market Orientation and Salesperson Customer Orientation: Interpersonal and Intrapersonal Influences on Customer Service and Retention in Business-to-Business Buyer-Seller Relationships. *Journal of Business Research*, *56*(4), 323–340. doi:10.1016/S0148-2963(02)00444-7

Kaplan, A. M., & Michael, H. (2010). Users of the world, unite! The challenges and opportunities of social media. *Business Horizons*, *53*(1), 59–68. doi:10.1016/j.bushor.2009.09.003

Kietzmann, J. H., Hermkens, K., McCarthy, I. P., & Silvestre, B. S. (2011). Social media? Get serious! Understanding the functional building blocks of social media. *Business Horizons*, *54*(3), 241–251. doi:10.1016/j.bushor.2011.01.005

Kimmel, A. J., & Kitchen, P. J. (2014). Word of mouth and social media. *Journal of Marketing Communications*, *20*(1-2), 2–4. doi:10.1080/13527266.2013.865868

Latane, B. (1981). The Psychology of Social Impact. *The American Psychologist*, *36*(4), 343–356. doi:10.1037/0003-066X.36.4.343

Marshall, G. W., Moncrieff, W. C., & John, M. (2012). Revolution in Sales: The impact of Social media and Related technology on the Selling environment. *Journal of Personal Selling & Sales Management*, *32*(3), 349–363. doi:10.2753/PSS0885-3134320305

Metzger, M. J., Flanagin, A. J., & Medders, R. B. (2010). Social and heuristic approaches to credibility evaluation online. *Journal of Communication*, *60*(3), 413–439. doi:10.1111/j.1460-2466.2010.01488.x

Mir, I., & Zaheer, A. (2012). Verification of social impact theory claims in social media context. *Journal of Internet Banking and Commerce*, *17*(1), 1–15.

Muntinga, D. G., Moorman, M., & Smit, E. G. (2011). Introducing COBRAs: Exploring motivations for brand-related social media use. *International Journal of Advertising*, *30*(1), 13–46. doi:10.2501/IJA-30-1-013-046

Nowak, A., Szamrejand, J., & Latane, B. (1990). From Private Attitude to Public Opinion: A Dynamic Theory of Social Impact. *Psychological Review*, *97*(3), 362–376. doi:10.1037/0033-295X.97.3.362

Nunnaly, J. (1978). *Psychometric theory*. New York: McGraw-Hill.

O'Reilly, K., & Marx, S. (2011). How young, technical consumers assess online WOM credibility. *Qualitative Market Research*, *14*(4), 330–359. doi:10.1108/13522751111163191

Pjero, E., & Kercini, D. (2015). Social media and consumer behavior – How does it works in albania reality? *Academic Journal of Interdisciplinary Studies*, *4*(3), 141–146.

Ridings, C.M. & Gefen, D. (2004). Virtual community attraction: why people hang out online. *Journal of Computer-Mediated Communication, 10*(1), 1-10.

Singh, S. (2016). Role of Social Media Marketing Strategies on Customer Perception. *Anveshak-International Journal Of Management*, *5*(2), 27–41. doi:10.15410/aijm/2016/v5i2/100695

Sun, S.-Y., Ju, T. L., Chumg, H.-F., Wu, C.-Y., & Chao, P.-J. (2009). Influence on Willingness of Virtual Community's Knowledge Sharing: Based on Social Capital Theory and Habitual Domain. *World Academy of Science, Engineering and Technology*, *53*, 142–149.

Swedowsky, M. (2009). *Improving Customer Experience by Listening and Responding to Social Media*. Retrieved September 8, 2013, from http://blog.nielsen.com/nielsenwire/consumer/improving-customer-experienceby-listening- and-responding-to-social-media/

Toor, A., Husnain, M., & Hussain, T. (2017). The Impact of Social Network Marketing on Consumer Purchase Intention in Pakistan: Consumer Engagement as a Mediator. *Asian Journal of Business and Accounting, 10*(1), 167–199.

Trusov, M., Bucklin, R., & Pauwels, K. (2009). Effects of Word-of-Mouth Versus Traditional Marketing: Findings from an Internet Social Networking Site. *Journal of Marketing, 73*(September), 90–102. doi:10.1509/jmkg.73.5.90

Vinerean, S., Cetina, I., & Tichindelean, M. (2013). The effects of social media marketing on online consumer behavior. *International Journal of Business and Management, 8*(14), 66–69. doi:10.5539/ijbm.v8n14p66

Williams, K. D., & Williams, K. B. (1989). Impact of source strength on two compliance techniques. *Basic and Applied Social Psychology, 10*(2), 149–159. doi:10.120715324834basp1002_5

Yang, T. (2012). The decision behaviour of Facebook users. *Journal of Computer Information Systems, 52*(3), 50–59.

Chapter 7

The Digital Campfire:
An Ontology of Interactive Digital Storytelling

Jouni Smed
University of Turku, Finland

Tomi "bgt" Suovuo
University of Turku, Finland

Natasha Trygg
University of Turku, Finland

Petter Skult
Åbo Akademi University, Finland

Harri Hakonen
Independent Researcher, Finland

ABSTRACT

Interactive digital storytelling (IDS) allows a human user to become an active part in a story and to affect how the story unfolds. To understand IDS systems, we need to consider the partakers present in them as well as their roles and interconnections. In this chapter, the authors discern four partaking entities—interactor, author, developer, and storyworld—and describe both their affiliated sub-entities as well as their relationship to one another. Based on both reviewing relevant literature and analyzing existing IDS systems, the ontology presented here provides a cohesive view into the current state of both theoretical and practical research.

DOI: 10.4018/978-1-5225-5715-9.ch007

INTRODUCTION

Counting from Brenda Laurel's (1986) doctoral dissertation three decades of research have gradually deepened the understanding of the partakers and their interconnections in interactive digital storytelling (IDS). The advances have been brought forward by both theoretical work – such as Brenda Laurel's *Computers as Theatre* (1991) and Janet Murray's *Hamlet on the Holodeck* (1997) or the doctoral theses by Michael Mateas (2002), Mark Riedl (2004), Sandy Louchart (2007) and Ernest Adams (2013) – and pioneering IDS systems such as *Façade* (Mateas & Stern, 2005), *Interactive Drama Engine* (Szilas *et al.*, 2007) and *ASAPS* (Koenitz & Chu, 2012). Nevertheless, despite these developments IDS still lacks a clear ontology and a common terminology (Koenitz, 2014).

The traditional division has recognized the roles of an author, characters and interactor (commonly used terms are also 'user' or 'player'). In this triad, the author is usually seen as the "creative force" defining the mechanics and content of the IDS system, and both the interactor and the characters then co-create each individual story instance within the IDS system's limits. However, this perspective focuses on the flow of the story, cloaking the crafting aspect of the story elements. Behind – or rather mixed with – the author is the developer of the IDS system who is responsible for the design of the software running the system, whereas the author is responsible for the content of the system. The developer and author together (which we jointly refer as 'providers') generate a semi-autonomous storyworld, the fourth partaker, where the interactor enters to experience the story instance. This is not unlike, for instance, Roman Jakobson's (1960, p. 353) notion of the functions of (verbal) communication, in which he emphasises the triad of addresser, message and addressee (i.e., sender, message and receiver) around which all communication is built.

The role of the character – albeit vital for the experience of an IDS system – is a part of a storyworld, which also includes props, scenes and events. One could argue that props are inanimate objects, which can be used in the storyworld, and events cause changes launched by fulfilling some criteria. Characters combine these two properties: they are both objects and agents of change. Scenes are the surroundings which the props and characters inhabit and where the events and characters can affect.

To summarize the ontology of IDS systems has four distinct partakers (see Figure 1):

- Interactor,
- Author,
- Developer, and
- Storyworld.

Figure 1. The ontology of IDS systems discerning the partaker roles of interactor, author, developer, and storyworld, and their corresponding sub-elements

In this chapter, we will go one-by-one through each partaker and the sub-entities connected to them, describing issues related to them and their connections to one another. It is the aim of this chapter to give a clear understanding of the roles, rights and responsibilities of everyone involved in the IDS activity. As in more traditional movie industry, the movie as an entity can be dissected into the target audience reaction, producer's decisions, director's choices, and actors' performances, etc., we present an equivalent top layer for similar dissection for IDSs

INTERACTOR

The interactor in IDS has a counterpart in traditional storytelling settings such as the reader of written stories or the audience of concerts and theatrical plays. However, instead of being a passive recipient of the story, the interactor is required to participate actively in the creation of the experience. This means making decisions that affect on how the story unfolds.

As Ernest Adams (2013) describes, the interactor undertakes an agreement to comply to the story, since – presumably – there is a reason why the author is leading the interactor through the story (Perlin, 2005). Moreover, we currently see in many examples from mainstream games that the author designs the gaming experience with aim to force its player to change the course of the game or, at least, to give an illusion of freedom in creating own story(world). Yet, if the player inclines to proceed in the game as the author may have intended (more linear storytelling approach), the author provides tools with which game can be played without making much of individual alterations to the story(world). Alternatively, the author can give branching story-choices which, at the end, lead to the same ending or a finite set of alternative endings.

Agency

Because of the interactivity, the interactor is not just a passive observer of choices made by the characters in the story, but the interactor makes choices for the story as well. It can be a choice of general plot direction every now and then or the choice of timing when to leap to cross a chasm or the real-time choices of what action to take as a character in the story. If the story keeps on proceeding the same way regardless if the interactor makes any choice, the story ceases to be interactive. A bare minimum of interactivity is that the story will advance only if the interactor takes an action such as pressing a button. One could, therefore, argue that a system, where one has the option to pause the story and continue it at will, is, in principle, a minimal IDS system.

For an IDS system to be genuinely interactive, the interactor's choices should affect the direction of the unfolding story. This *agency* "is the satisfying power to take meaningful action and see the results of our decisions and choices" (Murray, 1997, p. 126) and it is the distinctive experience that an interactor has in an IDS system. With this in mind, having a freedom of choice – choosing a certain path or viewpoint in which story will progress – can be considered as a minimum requirement for true IDS. Early examples of this are interactive films, for example, *Kinoautomat* (Kalas & Činčera, 1967) and *I'm Your Man* (Franzblau & Bejan, 1992) and interactive television programmes such as *Mörderische Entscheidung* (Schmidt & Hirshbiegel, 1991) and *D-dag* (Ehrhardt *et al.*, 1999).

Agency is facilitated by the developer and provided for by the author. The developer creates the interface through which the interactor can make choices, and the author defines what choices the plot allows and how they can affect it. The real depth of agency is relative to the level of narrative influence. Seymour Chatman (1978) divides the parts of the story plot into *kernel* elements that are fundamental to the story and *satellite* elements that are optional to the story. Table 1 illustrates the possibilities of affecting the kernels and satellites (Aarseth, 2012):

- If the interactor cannot affect the kernels nor the satellites, agency is shallow and the story will reduce into a linear story that will take the same course in all instances.
- If the interactor can influence the satellites, we have a structure typical to linear games.
- If the interactor has the liberty to choose the kernels from a set of alternatives but has no influence on the satellites, we have a non-linear story (e.g., hyperfiction).
- If the interactor can choose the kernels and can influence the satellites, we have deep agency, for example, in the form a quest game.

- If the interactor can influence both the kernels and the satellites, we have a pure game (e.g., chess).

One should note, however, that the agency experienced by the interactor is not necessarily in relation to the real depth of agency. The interactor may be given a deep sense of agency without giving them any real agency at all – similarly to the Eliza effect where a system appears to be more complex than it is. Conversely, the interactor may feel like their actions have no influence to the game whatsoever, even if the story mechanics in the background are thoroughly affected by each action the interactor takes. Noah Wardrip-Fruin (2009) calls this case, where a system fails to represent its internal richness, the Tale-Spin effect. He also describes the ideal situation, the SimCity effect, where a system enables the interactor to build an understanding of its (complex) internal structure.

Multiple Interactors

It is a different experience to listen to music in a concert than to have someone to play music to you alone. Also, watching a movie alone or with someone else are different experiences. The live audience interacts with each other both directly and indirectly through the performers.

IDS systems, however, are often intentionally designed for single interactors. If we allow multiple interactors in an IDS system, we must also prepare for conflicts in the design. In distributed databases, the conflict is about maintaining consistency so that all users have the same view to the shared data. In multiplayer online games, the conflict is about maintaining both consistency and responsiveness so that all players have a prompt access to relatively reliable game data (Smed & Hakonen, 2017, pp. 256–258). An IDS system with multiple interactors takes the conflict to a new level, because we have to create interwoven stories that are, at the same time, consistent, responsive and compelling. A compelling storyworld requires that the events are dramatically interesting to all the interactors.

Table 1. Influencing the kernels and satellites (Aarseth, 2012)

Kernel influence	Satellite influence not possible	Satellite influence possible
no influence	linear story	linear game
choose from alternatives	non-linear story	quest game
full influence	n/a	pure game

There are three questions that multi-interactor IDS systems have to address. The first and crucial challenge is how can we ensure that all interactors will stay in the focus of the story. This *too-many-heroes* problem asks how can we tell a story that would be compelling to multiple main characters (Smed, 2014). Most massive multiplayer online games ignore this problem and offer the same story for all players. For example, *World of Warcraft* (Blizzard Entertainment, 2004) allows everyone to take turns in killing the Lich King and saving the world. However, if we want to solve this problem, every character cannot be a hero but someone also has to do the mundane work – even if we are in a storyworld. At the heart of this problem is that each human-controlled character (i.e., hero) needs a group of computer-controlled characters (i.e., extras) to support them. Therefore, one solution is that each new interactor brings along also new supporting characters. Another approach is to limit artificially the number of interactors in the storyworld so that we can provide each of them with a meaningful story.

The second challenge is about *persistency*: if the storyworld is persistent, how do we handle interactors entering and exiting at any time? Multiplayer online games have to solve the same problem, but in an IDS system we have to consider also the on-going stories and the presence of interactors. For example, let us think about what happens when an interactor logs out. One possibility is that the interactor's character just vanishes from the storyworld, which is inconvenient and not believable unless it is included in the storyworld's internal logic. Second possibility is that the interactor's character becomes a computer-controlled character until the interactor logs back in. The problem is now how to guarantee that something extraordinary does not happen to the character in the meanwhile. When the interactor is not present, the character cannot be subjected to big plot twists. Naturally, we can present a recap of the events that have taken place during the interactor's absence upon returning. Third possibility is that the interactor gives tactical (or even strategic) level instructions to the character to follow during the absence (e.g., "try to befriend this other character", "stay home and do not answer the phone", or "be happy and active") (Smed & Hakonen, 2017, p. 284). However, many interactors might find this kind of a loss of control, even if it is only temporary, intrusive and confusing.

The third challenge is *cheating* in a storyworld and its implications. Apart from technical cheating such as hacking the software, this is a about what belongs to the agreement the interactors are committed to. Cheating means achieving the goal by breaking the rules, but what are the goal and rules in a storyworld? Cheating that takes place inside the storyworld is just a part of the story, since every action within the storyworld – no matter how civil or rude – are part of the experience and should be valid. This kind of cheating can be called managed or explicitly possible.

However, cheating that is not comprehended as a part of the interactors' agreement may ruin the experience, depending on if the cheat becomes accepted as a way to broaden the conflict aspect of the storyworld. That is, the agreement may evolve, with a mutual approval. Nonetheless, the only authority to cheat is the author – and that should always be allowed.

Representation

In order to be understood, the story must be concretized into a representation. This representation can be anything that conveys the experience of the story to the interactor and, at the same time, serves as a means for the author to express the story and vary it to reflect the interactor's reactions. With traditional medium the representation is typically fixed: paintings are visual, music is auditory, theatre plays are watched in theatres. With IDS, not only the representation may be multimodal but it can also be interactive. The interactor may be able to choose not to have audio input at all but, for example, rather to have the speech and other auditory cues through subtitles. A game can allow the player to move around in the physical world with the game device to control the game, or instead the player may choose to control the game with a joystick and simulate the motion in the physical world.

In the case of visual representations, visual storytelling possibly has the most significant effect on human reception. Seeing and reacting on an image has much more success in provoking senses and being memorized longer than any other verbal or written information. In this matter, Erwin Panofsky (2003, pp. 306–310) states that the steps of understanding correspond to the forms of knowledge which presuppose historical experience.

Early semiotic work is known as connoisseurship, which is related to the interpretation of signs of authenticity and authorship. The author holds the origin of the work of art and aims to characterize the existence, circulation and discourses within a storyworld. The author definitely beholds the attendance of certain events within the created world, along with changes, distortions and their various modifications. Lead by the author's thought with conscious or unconscious desires, the contradictions can resolve in relations to the others creating a specific meaning which is the interactors' to find.

In visual representations, parts of the field are open for submitting the order of values in the context of the represented objects with potential signs. The changing nature of image–sign relationship is essential subject in Meyer Schapiro's (1969) view. One could say that semiotic approaches and theorems about each aspect of visual representation (an artwork, a sentence, a word, a letter) are the matters in discussion of interactive storytelling in the digital era (Merleau-Ponty, 1964, p. 58).

Unlike verbal, the written text is under continuous transformational process of perception since each generation has own viewpoints and attitudes towards collected written material. That brings changes and movements in elements and makes differences in interrelations of the characteristics, which belong to the same elements of the visual representations. Artwork stands for the subject, behind the work stands the sender whose expression stays within the spectators. Observer's interpretation forms the basis of the common meaning generated by the interaction between visual and verbal discourse. We can see clearly this approach in poststructuralism where the study of interpretation aims to the balance of taking a look at the past as the act of construction. According to Gadamer (2004), the true power of visual representation is in its ability to shape the observer's understanding from what appearance suggests – observer can receive several different stories, or even several different aspects of the same story, by observing one visual allegory. The power of visual representation is in its ability to adjust the consciousness of the observer in which processes the idea of an artwork is crucial aspect of the concept which its appearance suggests.

The role of art has always been a double to the real world, being compared and evaluated to how real does it feel, or, in other words, how faithful it is to our senses of what real represents to an individual. Like any other form of art, game design and visual representations of the either fictional or historical elements, can be examined by using post-modern art theories. IDS systems (as well as digital games in general) provide a new medium of expression where the interactor does not regard them as entertaining platforms but digital environments for gaining new experiences.

AUTHOR

In conventional storytelling, the author's role is decisive in creating the presented story to the spectators, whereas in interactive storytelling the author's role is reduced (see Figure 2). Let us first consider non-interactive storytelling forms such as traditional books and movies. In this case, we have a single author – the writer or the director – composing a story to an audience, and the only interaction in storytelling happens before the story set down to its published form. For the spectators, the presented story is always the same but everyone makes their own experienced story out of it. In reality, the case is typically more complex and interactive than that. Books are processed by writers who receive feedback from editors and advance readers before the book is published. Movies are conducted by the directors in a complex process involving actors, cameramen and many others – and the director might not even get the last word but the film can go through a recut based on the reactions from test audiences. Nevertheless, on a higher level of abstraction, these teams can be considered collectively as the author.

Figure 2. Author's relation to the audience in conventional and in interactive storytelling. In IDS, the traditional role of author is divided by the providers (author and developer).

In contrast to this limited, almost one-directional and non-interactive storytelling, IDS systems redefine the role of the author and the whole authoring process. Murray (1997, pp. 185–213) anticipates the coming of the cyberbard, which would utilize the properties of digital media and create procedurally multiform stories open to collaborative participation. Murray further claims that even densely plotted works like the *Iliad* and the *Odyssey* were collective efforts of a highly formulaic oral storytelling system. This bardic system is conservative, focusing on the underlying patterns where a particular performance can be created. In IDS, the underlying patterns are provided by the developer and the author taking the role of Murray's cyberbard to create the content of the storyworld.

Authoring

In an IDS system, authoring is delivering content for somebody else's experience, which means that the author defines actions, states and events (Spierling, 2009; Spierling & Szilas, 2009). Consequently, one can ask whether 'author' still has authority in such a setup or should we refer the author as a 'designer' (Adams, 2013) or a 'narrative designer', which is now getting a more common professional role in commercial game development. We rather see the author in a position of shared authority together with the developer. Together they are the *providers* of the storyworld for the interactor.

At the core of authoring lies the *narrative paradox* – to which we will come back later when discussing about the developer – because the author cannot expect the interactor to make the right decision at the right moment or in the right place (Louchart & Aylett, 2005). For this reason, the author's role is to write interesting characters and rely on their ability to interact with one another (the interactor can be considered to be an autonomous actor as well). The author must, therefore, be attentive to the interactor's inner state.

The challenges for authoring include (Spierling, 2009; Spierling & Szilas, 2009; Aylett & Louchart, 2011):

- Due to the medium's immaturity the tools used for authoring often show the underlying software solutions and the line between storyworld and the story engine can be blurry. For example, the content might depend on the run-time system architecture.

- As the amount of required content increases the more complex the storyworld gets. This implies that authoring might not be a single author's task but should support multiple authors. Some of the authors can be responsible for the visual content, others for authoring the audio world (e.g., music, sound effects, voice-overs), or for simple text (e.g., in-game letters or emails between characters, background lore), or for putting the disparate pieces of authored texts together into cutscenes and in-game cinematics. The author, and authoring, is often a collective effort.

- The usability of IDS systems requires that the story-related structures are presented at a suitable abstraction level for the authors. Narrowly formatted and constrained authoring mechanisms limit the author's possibilities to reduce human affairs into logical models. On the other hand, to support the author to utilize the potential of a story engine requires inspiring examples and prototypes as a study material.

Designing

In IDS authoring, the mere narrative is typically not sufficient since the storyworld has to also contain visuals and sounds. The author is mainly responsible for the first impression the IDS system gives, and generally the visible and audible design. Ideally, the audiovisual representation of the storyworld should guide the interactor through the story, and the design of an IDS system should be both functional and aesthetic. Successful authoring creates memorable experiences. The developer, on the other hand, is responsible of the interaction design, which requires mediating between the interactor and the author, delivering the content from the author to the interactor, and listening to the choices made by the interactor and comparing them to the story design described by the author.

One criteria of an engaging story is the extent to which the interactor feels connected to it: how difficult it is to put the book down or how tempting it is to play "one more turn". As with all forms of entertainment, immersion is often the key, and Marie-Laure Ryan (2001, p. 103) calls effective immersion 'recentering'. This means transporting the reader (interactor in our terminology) into the fictional world and making that fictional world the new center of understanding for the interactor: a successful recentering allows for the suspension of disbelief, a vital aspect of immersion. As stated by Ermi and Mäyrä (2005), immersion in games has three dimensions: sensory-based immersion, imaginative immersion, and challenge-based immersion. This model suits all types of IDS. The *sensory-based immersion* entails the aesthetics of the system. This is provided by the audiovisual content provided by the author as well as the IDS system's high fidelity and timing provided by the developer. The *imaginative immersion* entails the attractiveness of the story, which is mainly provided for by the author, who designs the story content. Through the affordance of interactivity, this dimension is even more affected by the system built by the developer that is responsible of keeping the audience's interest. The *challenge-based immersion* entails the experience of agency. A game that is difficult but still surpassable and has an intuitive control scheme, has a quality of keeping the interactor's attention.

DEVELOPER

Developer is an elemental role to the digital variety of interactive storytelling. When telling stories by writing books or painting images, it can essentially be expected that the author self-masters the tools and mechanisms involved. Of course, most writers and painters do not make their own ink and paper or colours and canvas. If you consider this production as a part of the storytelling process, the people responsible on these tasks could be considered as 'developers' in the same sense as used in our ontology. Yet, it is more typical in the digital media that although the author purchases a computer as a tool (like they would purchase ink and paper as a book writer), there is a need for a special scribe called "the programmer" (or as our ontology generalizes it more, 'the developer') who masters the techniques of transferring the author's vision into a functional digital artefact.

Selecting a Narrative Approach

The developer's task is to select a strategy on how to handle the *narrative paradox*, which happens when the "pre-authored plot structure conflicts with the freedom of action and interaction characteristics of the medium of real-time interactive graphical environment" (Aylett & Louchart, 2007). This creates a tension between the interactor's freedom and well-formed stories (Adams, 2013), which can be seen in two ways (Louchart, 2005):

1. The plot constraints the interactor's freedom, and
2. Interactive freedom affects the unfolding of the story.

Simply put, the more freedom the interactor has, the less control author has, and vice versa.

Here, the developer can choose from two opposite approaches: the author-centric and the character-centric. The *author-centric approach* (also known as explicit authoring, top-down or plot-centric approach) models the creative process of a human author. The system includes a proxy of the author, the drama manager, which controls the events and characters of the storyworld. The *character-centric approach* (also known as emergent narrative, bottom-up or implicit creation) focuses on autonomous characters and modelling the mental factors that affect how the characters act. The story emerges from the characters' decisions and interaction.

To compare the two approaches Riedl (2004, pp. 12–14) proposes two measures:

1.   **Plot Coherence:** The perception that the main events of a story are causally relevant to the outcome of the story.
2.   **Character Believability:** The perception that the events of a story are reasonably motivated by the beliefs, desires and goals of the characters.

Clearly, author-centric approach allows us to have a strong plot coherence as a result of the drama manager's influence. The downside is, however, that the character believability weakens when the actions of the characters seem to be compelled to follow the author's will (Aylett *et al.*, 2011). The problem is then finding subtlety so that the influence does not feel too forced upon the user. In implementation, the main concern is that an IDS system must observe the reactions of the interactor as well as the situation in the storyworld to recognize what pattern fits the current situation.

Conversely, character-centric approach has (and requires) a strong character believability. This means that the plot coherence is weaker, because the story emerges bottom up based on the characters. Although the idea of emergent narrative of the character-centric approach seems to solve the narrative paradox, it is unlikely that it is enough for implementing a satisfying IDS system (Aylett *et al.*, 2011). Realistic actions are not necessarily dramatically interesting, if the characters have no dramatic intelligence. Therefore, the argument is that the author's presence is necessary, because without the author's artistic control we would end up having the chaos of everyday life.

This leads to idea that the approaches can be combined to a hybrid model, where a character proposes a set of possible actions to a drama manager, which selects the dramatically best alternative (Weallans *et al.*, 2012). Here, the drama manager is no longer pushing the characters to follow its lead but supports their decision-making.

Relation to Author

The relationship between the author and developer is close. As IDS systems are not yet mature enough, they are being developed and modified according to the feedback from the authors as well as interactors. However, separating the roles is vital and, clearly, there is a dependency where the author is using the IDS system created by the developer. Figure 3 illustrates the boundaries of authoring and emphasizes that the developer and the author should not be the same. The developer's responsibility is the runtime engine that enables the performance of the agents' autonomous or semiautonomous behaviour, whereas the author's storyworld constitutes the actual "content".

Relation to Interactor

A modern trend in IDS, and particularly in games, is providing support to modifications (called 'mods') made by the player community. This places the members of the community in the role of the developers. In other words, the role of the developer does not only include people from the company who publishes the game, but also from those who traditionally are considered as the audience. These modifications can extend the functionality or mechanics, which can even transform original system into something that the original developer had not intended.

The mods can involve the mechanisms that the main authors use to put the story into the game. Thus, through mods, the community of interactors (the fandom) may become authors for storyworlds in the IDS system. These community created storyworlds can even be the basis for the business idea like in the IDS systems *Episode* (Episode Interactive, 2013) and the now-defunct *Versu* (Versu, 2013).

Figure 3. The boundaries between the interactor, author, and developer (Spierling & Szilas, 2009). The IDS artefact consists of the pieces of the runtime engine the providers have created.

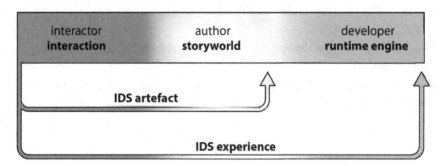

STORYWORLD

The storyworld is the artefact provided by the authors and developers for the interactor to experience. Usually the storyworld is designed by the providers and only experienced by the interactors. In the co-design paradigm, essentially everyone involved with the storytelling are also involved with designing the storyworld, which is the case with tabletop games and live action role-playing games.

The storyworld has various elements with different levels of autonomy. *Characters* are computer-controlled entities that are represented as avatars in the storyworld and interact with the interactor. In non-digital interactive storytelling, such as traditional role-playing games, characters are controlled by a human dungeon master. Interactors and characters can use *props* to act and to interact in the storyworld. The developments in a story can raise also due to *events* that are usually set by the author and triggered by the interactor. The storyworld also comprises of various *scenes* where the story takes place.

Generally, all stories have characters of some sort, and the stories consist mostly of their relations and interactions between them. In classical narratology, the most important questions regarding the characters in the story directly relate to the communicational content of what is being told: who is the teller and who is it being told to. Gérard Genette (1980, p. 189) uses the term 'focalization', often confused with the more common 'point of view', to discuss the various ways the author and the characters in the text interact with the reader.

Typically stories are focalized in a very specific way: Genette speaks of 'zero', 'internal' and 'external' focalization (Genette, 1980, pp.189–190). In zero focalization, a form of 'omniscient' narration, the narrator knows more than the characters know, and thus so do the readers. In 'internal' focalization, the narrator and the character are one and the same, and everything the narrator/character knows the reader knows. Finally, 'external' focalization allows for a situation where the character knows more than the narrator: the narrator may only objectively describe the character's actions and external appearance, but not their internal world. There is of course no reason not to mix these within the same narrative, nor any reason to remain within a single character for the whole duration. Focalization in games follows roughly similar patterns, although, unlike other forms of narrative media, 'character' is not always equally central. For example, simulation or sports games may well be populated by multiple actors, even named and voiced actors, but to call them characters might be giving them too much credit, as their internal worlds are often entirely irrelevant to the game being played or the story being told; they are 'externally' focalized in the sense that their appearance and actions are described, but there is no assumption that were one to suddenly inhabit (internal focalization) their minds, there would actually be anything in there, so to speak.

There are, however, plenty of video games which offer variations on both internal and zero focalization, as well as 'proper' external focalization (where one can safely assume a character has a story that we are simply not privy to). First or third person games, for example, might feature internal dialogue or commentary on the environment that is only uttered out loud because the player needs to have access to the character's thoughts – such as in *Outlast 2* (Red Barrels, 2017) or *BioShock Infinite* (Irrational Games, 2013). But first person games can also be externally focalized, as in the first *BioShock* (2K Boston & 2K Australia, 2007) or *Half-Life* (Valve, 1998), where we have no access to the internal world of the protagonist we are inhabiting. Zero focalized games often feature narrators, such as in *Bastion* (Supergiant Games, 2011), that freely inform the player what the characters are thinking, or otherwise feature in their user interface elements that let the player know what the character is feeling, as in, for example, *The Sims* (Maxis, 2000) or *RimWorld* (Ludeon Studios, 2016) that have detailed statistic on the emotional state of the multiple characters the player can control.

The most common form of character focalization in games is, however, inherited from film (i.e., the external). Much like in cinema, we can only assume what the actors are thinking or feeling, based on their animations or dialogue lines – and this includes the so-called 'player' character (if there is one).

Finally, we cannot omit the importance of the developer's decisions, which set the "natural laws" to the storyworld. The most important of them is the choice of narrative approach – author-centric, character-centric or hybrid – which crucially affects how the story is generated in the system and how the storyworld can be constructed.

THE STORY

Following Jakobson (1960), the 'story' in IDS would be the 'message', in other words, what lies between the author and the interactor or what the author is attempting to present to the interactor through the storyworld and its characters. In IDS, the story is a two-fold concept. First, there is the author's idea of the story. The author and developer create the storyworld to deliver an experience for the interactor. They put in the theme and, depending on the type of story they wish to tell, might put in everything from a singular path with rigid story points that must be followed in a specific order, to a sprawling possibility space filled with individual, hand-crafted or generated 'storylets' (to borrow a term from the StoryNexus storytelling system by Failbetter Games) that can be experienced in any order, to anything in-between. Second, individual stories begin to diverge and arise from this Platonic ideal of a story as a result of the interactor's experiences. The extent to which these individual stories are permitted to diverge from the intended experience depends on the intent of the developer and the author; although divergent experiences can also be manufactured by industrious interactors entirely outside the realm of the 'intended', for example, speedrunners will purposefully break games in order to finish them quickly and thus often entirely subvert the intended narrative by skipping over whole sections or glitching through levels. Following in the footsteps of Jakobson, we find that the role of the 'message' (i.e., the story) in an IDS can vary considerably. Jakobson described six different functions of communication, which describe the reasons and intention of why we communicate, each with a focus on a different part of the addresser–message–addressee scheme (see Figure 4).

Most of these functions have to do with everyday verbal communication, but are also surprisingly easy to apply to the role and purpose of 'story' in an interactive experience. The *emotive function* (Jakobson, 1960, p. 354) refers to communication for the sake of (on the behalf of the sender) expressing some particular emotion, such as anger or interest, and is usually non-verbal or interjective ('Tut tut!' to express disappointment): story-wise, this might be the game telling the interactor they have done something correctly or wrong, the game reacting in some way to the player's actions. The *conative function* (Jakobson, 1960, p. 355) on the other hand focuses on the receiver, the addressee, and is best expressed in the imperative: communication occurs in order to give the receiver some form of instruction; for example, consider tutorials in many games, which are often clothed in some form of 'story'. The *referential function* (Jakobson, 1960, p. 55) has to do with context: where is the communication taking place. Particularly in games low on other story content, this is one of the prevailing places where story is imparted, as the referential function would be how we can tell that this game takes place in, say, a candy kingdom à la *Candy Crush* (King, 2012). There are also phatic and metalingual functions: *phatic functions* are simply used to make sure the channel of communication is functioning, or then to prolong communication without offering any new information – this might be like a character in the game asking you if you are still playing, or the player starting to whistle at random during a prolonged idle period. The *metalingual* (or

Figure 4. Jakobson's (1960) addresser–message–addressee scheme with emotive, conative, referential, poetic, phatic, and multilingual functions

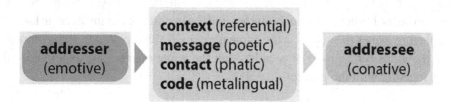

glossing) *function* (Jakobson, 1960, p. 356) is used to discuss the code (i.e., the language) itself, typically by asking someone to repeat something they said, or making sure they are following: within IDS this might manifest through a tutorial prompt or an advisor character commenting on some aspect of the game trying to make sure they are following along. The final function of language is the *poetic function* (Jakobson, 1960, p. 356), which is communication focused on the message itself. This, naturally, is how the most interesting stories are told, and is the purvey of all forms of narrative media.

IDS is, out of all forms of narrative art, unique in that many games make do with the simpler functions of communication when they tell their story; they may simply employ 'story' in order to communicate a tutorial or to give the interactor a referential frame within which they can perform various actions that are, otherwise, entirely divorced from any kind of deeper storytelling structures (matching three candies of the same type together…). For example, many simple games will have a theme (referential), a few characters that offer encouragement at success and failure (emotive) and advice when new gameplay elements are introduced (conative), as well as now and again prodding the player when they are idle or when too long a pause in the action occurs (phatic), without offering in any way a cohesive or intentional story as we typically think of it (the poetic function).

CONCLUDING REMARKS

We have presented here an ontology of IDS. In the era of digital culture, it has become more important to recognize the sources of interaction and see how they affect all the actors in the presented ontology. Instead of being just a passive audience, the interactors have an active role in creating the story. Also, the storyworld includes many computer-controlled mechanisms – most notable intelligent characters – that provide their input into the mixture.

As in any classification, drawing the lines is not always easy or unambiguous. For instance, the border between the roles of developer and the author can be very vague. When an IDS system is created by a single individual, this person is then, naturally, both the author and the developer. In larger organizations, people have mixed roles: some may be exclusively developers, creating the infrastructure only, and some may be purely authors focusing solely on the story. Another problematic area can be the border between the author and the interactor. For instance, in tabletop roleplaying games the author is the game master who has created the adventure and runs the game for a group of people – but the game master also experiences the story together with the players.

IDS is in a constant flux, because the technology is still far from mature. Nevertheless, we are seeing how the theoretical framework and understanding is getting firmer and more cohesive. Also, the IDS systems – albeit often developed for quite limited setups – are propelling the field forward. The most likely utilizer of this work will emerge from video games, which have, for a long time, shown a keen interest in including IDS in the game design. That is not to say that IDS will be used only for entertainment purposes, but it will show its potential in more serious applications ranging from physical and psychological well-being to teaching skills, sharing culture and increasing the understanding of the human condition.

Human beings crave for stories. That is why we want to tell them and hear them, affect them and be affected by them. IDS is the next step where the evolution of digital media is leading to – an era of a digital campfire.

REFERENCES

2K Boston & 2K Australia. (2007). *BioShock* [Software]. USA: 2K Games.

Aarseth, E. (2012). A narrative theory of games. In M.S El-Nasr, M. Consalvo, & S. Feiner (Eds.), *Proceedings of the International Conference on the Foundations of Digital Games* (pp. 129–133). New York, NY: ACM.

Adams, E. W. (2013). *Resolutions to Some Problems in Interactive Storytelling* (Doctoral dissertation). University of Teesside, Middlesbrough, UK.

Aylett, R., & Louchart, S. (2007). Being there: Participants and spectators in interactive narrative. In M. Cavazza & S. Donikian (Eds.), *Virtual Storytelling. Using Virtual Reality Technologies for Storytelling. Proceedings of the Fourth International Conference (ICVS2007)* (*Vol. 4871*, pp. 117–128). Berlin, Germany: Springer-Verlag. 10.1007/978-3-540-77039-8_10

Aylett, R., Louchart, S., & Weallans, A. (2011). Research in interactive drama environments, role-play and story-telling. In M. Si, D. Thue, E. André, J. Lester, J. Tanenbaum, & V. Zammitto (Eds.), *Interactive Storytelling: 4th International Conference on Interactive Digital Storytelling (ICIDS 2011)* (*Vol. 7069*, pp. 1–12). Berlin, Germany: Springer-Verlag. 10.1007/978-3-642-25289-1_1

Blizzard Entertainment. (2004). *World of Warcraft* [Software]. USA: Blizzard Entertainment.

Chatman, S. (1978). *Story and Discourse: Narrative Structures in Fiction and Film.* Ithaca, NY: Cornell University Press.

Ehrhardt, B. (Producer), Kragh-Jacobsen, S. (Director), Levring, K. (Director), Vinterberg, T. (Director), & von Trier, L. (Director). (1999, December 31). *D-dag* [Television broadcast]. Denmark: DR.

Episode Interactive. (2013). *Episode* [Software]. USA: Episode Interactive. Available from https://home.episodeinteractive.com/

Ermi, L. & Mäyrä, F. (2005). Fundamental components of the gameplay experience: Analysing immersion. *Worlds in Play, 37*(2).

Franzblau, B. (Producer), & Bejan, B. (Director). (1992). *I'm Your Man* [Motion picture]. USA: Interfilm Technologies.

Gadamer, H.-G. (2004). Truth and Method (2nd ed.). London, UK: Bloomsbury Academic.

Genette, G. (1980). *Narrative Discourse: An essay in method*. New York, NY: Cornell University Press.

Irrational Games. (2013). *BioShock Infinite* [Software]. USA: 2K Games.

Jakobson, R. (1960). Linguistics and poetics. In T. A. Sebeok (Ed.), *Style in Language* (pp. 350–377). Cambridge, MA: The MIT Press.

Kalas, L. (Producer), & Činčera, R. (Director). (1967). *Kinoautomat* [Motion picture]. Czechoslovakia.

King. (2012). *Candy Crush* [Software]. UK: King.

Koenitz, H. (2014). Five theses for interactive digital narrative. In A. M. C. Fernández-Vara, & D. Thue (Eds.), *Interactive Storytelling: 7th International Conference on Interactive Digital Storytelling (ICIDS 2014)* (*Vol. 8832*, pp. 134– 139). Berlin, Germany: Springer-Verlag.

Koenitz, H., & Chen, K.-J. (2012). Genres, structures and strategies in interactive digital narratives – analyzing a body of works created in ASAPS. In D. Oyarzun, F. Peinado, R. Young, & A. Gonzalo Méndez (Eds.), *Interactive Storytelling: 5th International Conference on Interactive Digital Storytelling (ICIDS 2012)* (*Vol. 7648*, pp. 84–95). Berlin, Germany: Springer-Verlag. 10.1007/978-3-642-34851-8_8

Laurel, B. (1986). *Toward the Design of a Computer-based Interactive Fantasy System* (Doctoral dissertation). Ohio State University, Columbus, OH.

Laurel, B. (1991). *Computers as Theatre*. Reading, MA: Addison-Wesley.

Louchart, S. (2007). *Emergent Narrative – Towards a Narrative Theory of Virtual Reality* (Doctoral dissertation). University of Salford, Salford, UK.

Louchart, S., & Aylett, R. (2005). Managing a non-linear scenario – a narrative evolution. In G. Subsol (Ed.), *Virtual Storytelling. Using Virtual Reality Technologies for Storytelling. Proceedings of the Third International Conference (ICVS 2005)* (*Vol. 3805*, pp. 148–157). Berlin, Germany: Springer-Verlag. 10.1007/11590361_17

Ludeon Studios. (2016). *RimWorld* [Software]. Canada: Ludeon Studios.

Mateas, M. (2002). *Interactive Drama, Art and Artificial Intelligence* (Doctoral dissertation). Carnegie Mellon University, Pittsburgh, PA.

Mateas, M., & Stern, A. (2002). *Architecture, authorial idioms and early observations of the interactive drama Façade.* Technical Report CMU-CS-02-198, School of Computer Science, Carnegie Mellon University, Pittsburgh, PA.

Mateas, M., & Stern, A. (2005). *Façade* [Software]. USA: Procedural Arts. Available from http://www.interactivestory.net/

Maxis. (2000). *The Sims* [Software]. USA: Electronic Arts.

Merleau-Ponty, M. (1964). *Signs* (R. McCleary, Trans.). Evanston, IL: Northwestern University Press.

Murray, J. H. (1997). *Hamlet on the Holodeck: The Future of Narrative in Cyberspace.* Cambridge, MA: The MIT Press.

Murray, J. H. (2012). *Inventing the Medium: Principles of Interactions Design as a Cultural Practice.* Cambridge, MA: The MIT Press.

Panofsky, E. (2003). *Iconography and Iconology: An Introduction to the Study of Renaissance Art.* Chicago, IL: University of Chicago Press.

Red Barrels. (2017). *Outlast 2* [Software]. Canada: Red Barrels.

Riedl, M. (2004). *Narrative Generation: Balancing Plot and Character* (Doctoral dissertation). North Carolina State University, Raleigh, NC.

Ryan, M.-L. (2001). *Narrative as Virtual Reality: Immersion and Interactivity in Literature and Electronic Media.* Baltimore, MD: Johns Hopkins University Press.

Schapiro, M. (1969). On Some Problems in the Semiotics of Visual Art: Field and Vehicle in Image-Signs. *Semiotica, I,* 223–242.

Schmidt, G. (Producer), & Hirschbiegel, O. (Director). (1991, December 15). *Mörderische Entscheidung* [Television broadcast]. Germany: Das Erste & ZDF.

Smed, J. (2014). Interactive storytelling: Approaches, applications, and aspirations. *International Journal of Virtual Communities and Social Networking*, 6(1), 22–34. doi:10.4018/ijvcsn.2014010102

Smed, J., & Hakonen, H. (2017). *Algorithms and Networking for Computer Games* (2nd ed.). Chichester, UK: John Wiley & Sons. doi:10.1002/9781119259770

Spierling, U. (2009). Conceiving interactive story events. In I.A. Iurgel, N. Zagalo, & P. Petta (Eds.), *Interactive Storytelling: Second Joint Conference on Interactive Digital Storytelling (ICIDS 2009)* (*Vol. 5915*, pp. 292–297). Berlin, Germany: Springer-Verlag.

Spierling, U., & Szilas, N. (2009). Authoring issues beyond tools. In I.A. Iurgel, N. Zagalo, & P. Petta (Eds.), *Interactive Storytelling: Second Joint Conference on Interactive Digital Storytelling (ICIDS 2009)* (*Vol. 5915*, pp. 50–61). Berlin, Germany: Springer-Verlag.

Supergiant Games. (2011). *Bastion* [Software]. USA: Warner Bros. Interactive Entertainment.

Szilas, N., Barles, J., & Kavakli, M. (2007). An implementation of real-time 3D interactive drama. *Computers in Entertainment*, 5(1), 5. doi:10.1145/1236224.1236233

Valve. (1998). *Half-Life* [Software]. USA: Sierra Studios.

Versu. (2013). *Versu* [Software]. USA: Versu. Available from https://versu.com/

Wardrip-Fruin, N. (2009). *Expressive Processing: Digital Fictions, Computer Games and Software Studies*. Cambridge, MA: The MIT Press.

Weallans, A., Louchart, S., & Aylett, R. (2012). Distributed drama management: Beyond double appraisal in emergent narrative. In D. Oyarzun, F. Peinado, R. Young, & A. Gonzalo Méndez (Eds.), *Interactive Storytelling: 5th International Conference on Interactive Digital Storytelling (ICIDS 2012)* (*Vol. 7648*, pp. 132–143). Berlin, Germany: Springer-Verlag. 10.1007/978-3-642-34851-8_13

KEY TERMS AND DEFINITIONS

Agency: The interactor's possibilities to make meaningful decisions in the storyworld and to see their outcomes.

Author: Responsible for creating the content of the storyworld using the tools and mechanisms provided by the developer. In contrast to traditional storytelling, the author has no direct control over the generated story.

Character: A computer-controlled entity inhabiting the storyworld usually represented by an avatar. Characters interact with one another and with the interactor, which affects the story being generated.

Developer: Responsible for creating the tools and underlying mechanisms of an IDS systems. The tools are used by the author to create the content of the storyworld. The developer and author together are called the providers.

Interactor: An active human participant in an IDS system, who has agency to influence how the generated story unfolds.

Narrative Paradox: The conflict between the interactor's freedom of choice (or agency) and the author's control over the storyworld. In extremes this means that if the author has a total control, the interactor has no agency, or if the author has no control, the storyworld is just a simulation.

Representation: Digital content (items, characters, environments, and other story elements) that provides IDS setting. The storyworld is generated by this content and serves the user with all necessary information for understanding the given surrounding, characters, quests, and other factors that shape the story progression.

Storyworld: A virtual world created by the providers (the author and the developer), populated by characters, props, scenes, and events, where the interactor can experience a story.

Chapter 8
Tweeting About Business and Society:
A Case Study of an Indian Woman CEO

Ashish Kumar Rathore
Indian Institute of Technology Delhi, India

Nikhil Tuli
Indian Institute of Technology Delhi, India

P. Vigneswara Ilavarasan
Indian Institute of Technology Delhi, India

ABSTRACT

This chapter examines the social media content posted by a woman Indian chief executive officer (CEO) on Twitter. The active involvement of CEO in communication activities influences the business effectiveness, performance, and standing of the business headed by her. Rstudio and Nvivo, two analytical tools, were used for different analysis such as tweets extraction and content analysis. The findings show the various themes in CEO communication which are categorized in different sectors in terms of her personal views (feelings and status updates), political views, and social concerns (ranging from education, women empowerment, governance, and policy support). The chapter extends the theoretical and empirical arguments for the importance of CEOs' social media communications. Finally, this research suggests that with a well-planned and strategic social media use, CEOs can create value for themselves and their businesses.

DOI: 10.4018/978-1-5225-5715-9.ch008

INTRODUCTION

Today people find themselves surrounded by multiple communication channels. The traditional mediums of communication like a newspaper, television are proven methods to disseminate information. There is a one-way sharing of information which lacks interpersonal capabilities ingrained in the way Internet lets us communicate (Steyn, 2004). Moreover, the interaction through social media tends to be more informal. It revolutionizes the way people shop, pay bills, communicates, etc. E-mail, blogs, social networking sites like Twitter, Facebook, and LinkedIn etc. could be described as some of these interactive media which make information available on fingertips (Rybalko & Seltzer, 2010; Rathore & Ilavarasan, 2018). In this way, social media can become the source of information to businesses (Rathore et al., 2016). The trend of social communication via twitter exploded in 2008-09 when various companies around the world started to experiment with the application (BRAND fog, 2014).

Though, slowly and with a lot of suspicions, the top level executives of the organizations have adopted this unique media into the personal medium of communication. They are directly linked to their associated businesses, so the information shared with the audience on social media is bound to be, at least implicitly, associated with the brand the senior executive represents. There is a strong and positive connection between CEOs communication quality and responsiveness because of their social media presence (Men, 2015). It affects the behaviour of external and internal stakeholders as well. As CEO represents the higher level corporate spokesperson, their participation in communication programme influences the different public relation activities and the organizational standing. CEO plays an important role by managing effective communication system which shapes the culture and the character of organization (Hutton et al., 2001). This reshaped structure of organization includes the various relationships among communities based on communication hierarchies (Men, 2014b).

It is amply clear that social existence and the communication made by CEO straddles the borderline between what could be personal and professional space. Unlike their predecessors, memo, meetings, press releases, conference calls, Twitter or to say any social media offers a unique and supposedly unfiltered and participatory access well as strengthen existing relationships with various stakeholders to CEOs (Karaduman, 2013). CEO and C-suite participation in social media mean good things for the business such as the organization's values, shape its reputation, and enhance its brand image (PwC Report, 2013). Further, it also helps their leadership respond better in times of crisis. Therefore, the analysis of CEO's personal communication is an enigmatic space (Porter et al., 2015). And from an academic standpoint, CEOs interacting online or tweeting on social media needs to be further explored.

This study examined the social media content through Twitter on cognitive and attitudinal aspects, particularly focusing on the use of Twitter by a woman Indian chief executive officer. The paper extends the theoretical and empirical arguments for the importance of her social media communications. In next section, a review of literature is discussed highlighting stats of CEOs presence on Twitter and their public relations.

REVIEW OF LITERATURE

CEOs Presence on Twitter

Global CEOs have embraced social media communication through the growth seems to be lagging far behind compared to the general public (Social CEO Report, 2013). The two social networks, Twitter and LinkedIn, though stand out in terms of respectable growth rate in the number of CEOs embracing the new space. Twitter is a micro-blogging social media platform which is used to obtain breaking news; communicate with friends, celebrities, and companies; follow the latest score of sporting events; etc. Users post "tweets" or mini-posts up to 140 characters in length via mobile texting, instant messaging, third-party applications, or the web (Rybalko & Seltzer, 2010). Unlike other social networking sites, "following" users on Twitter is not mutual. Other users have an option to follow members if their profiles are public, or they can ask permission to follow private member profiles. There are currently (5.6% CEOs on Twitter, which is definitely an improvement over last year's 3.6% CEOs. The report categorizes certain CEOs as active Twitter users on the basis of tweets in the last 100 days. The survey highlights that only 3.8% of the Fortune 500 CEOs are active Twitter users (Social CEO Report, 2013). Further, while Twitter's active global users are tweeting about twice per day, Fortune 500 CEOs who are active on Twitter are tweeting an average of 0.98 tweets per day (Social CEO Report, 2013).

It is important to understand the CEO mediated relationships in the space of collaborative interactions. The online content form CEOs are referred as open dialogues contributing to trust inauthenticity because they are alleged as being sociable and genuine (Men & Tsai, 2016). In most of the cases, CEOs are more likely good listeners, and less scripted and distant (Shandwick, 2012). The role of CEOs has become more critical in information seeking, building connections, and job-related benefits considering user needs, expectations, and gratification. Social media also offers a tool to make a real-time and spontaneous response in the context of public engagement (Waters et al., 2009). With social media presence, CEOs participate in ongoing discussion applying leadership strategy on a global scale. Social media features allow CEOs to engage with their all stakeholders in very authentic and informal manner (Men, 2015).

Need for CEOs to Embrace Twitter

The reputation of CEO in terms of personal attitudes influences various corporate investment and risk-taking decisions (Ahern & Dittmar, 2012; Borghesi et al., 2014; Resick et al, 2009). Due to sensitive reputation, CEO communication is considered as a crucial component of leadership demonstrating assertive and responsive content. Quality communication can be determined by various characteristics such as compassionate, friendly, sincere, understanding, and interested (Men, 2015). It helps in building an organizational reputation in a dynamic and competitive environment (Kim & Rhee, 2011). For quality content, it is a critical condition for CEO to make communication positively associated with openness and transparency (Men and Stacks, 2013). As a significant predictor, social media provides useful information in defining dialogue and interaction (Watkins; 2017). Social media platforms have changed the corporate communication dynamics in several major ways. First, while only elite journalists could produce news on corporate mistakes or wrongdoings in the past, now any Internet user can publicly discuss his negative experience about a company (Jameson, 2014). Second, social media enable direct interactions between individual customers and high profile corporate figures such as CEOs. Customers, who could access CEOs only through TV or magazine interviews in the past, now have a direct conversation channel with CEOs through platforms like Twitter. Third, social media expanded the scope of corporate communication in general. Corporate communication used to exist only between the public relations team and journalists, but now it exists virtually between any corporate personals and customers" (Park, 2011).

The senior executive of a company during his interaction with college students argued in favor of price discrimination based on the content usage of internet services (Alghawi et al., 2014). All this happened in a semi-formal environment and the executive left, satisfied with having put his points forward to the next generation. Unaware, the comments went online and there was a whole lot of discussion regarding the company's policies. The CEOs comments were being directly ascribed to the way company plans for the future. Wary of social media, CEO had no presence on any of the social media – Facebook, Twitter etc. There was no way to mention his side of the story and the company let discussions die down without any comment from the top leadership on social media (Jameson, 2014). This is a typical example of CEOs trying to think that social media won't have any ramifications on the existing organizational setup.

Public Relations Through Twitter

Social media presence of top executives of a company can go a long way in managing crisis communication. The study follows a conceptual model and tries to analyze the influence of medium and message on the recipient's perception of reputation, recipients' secondary crisis communications (e.g., sharing information and leaving a message) and reactions (e.g., willingness to boycott) (Schultza et al., 2011). The influence of the different media types i.e. newspaper, blogs, and twitter was analyzed separately. Also, the different modes of crisis communication were tested for their efficacy to reduce the existing negative sentiment. The top leadership must lead this march towards digitization and ensure an active presence on Twitter. A company or a top-level executive can't afford to mitigate the crisis by starting to tweet at the time of crisis. The engagement has to be long-term, and authentic in nature.

In any organization being a high-profile, CEO is the more visible entity to the public. Most of the online users tend to build a conversational relationship with such high-profile personae (Kim et al., 2016). To understand the CEO online personality might be useful to enhance the social relationship. For instance, in a crisis response, CEO personality can be emphasized to provide a better direction to practitioners. Social media presence of CEOs improves the reputation of organization (Shandwick, 2012). Therefore, CEOs are shifting their image distant to approachable and personable which cultivates value connections (Vidgen et al.,2013; Men & Tsai, 2016). To achieve highly interactive communication, CEOs make social media an ideal platform for actual dialogues. For successful crisis communication, it might, therefore, be important to address twitter users (Hwang, 2012). This acts as another stimulus for the businesses to show active presence on social media.

Social media seeks to blur the existing boundaries of a traditional organization since recent years have witnessed a number of companies and their top level employees embracing this powerful medium to share information with all stakeholders (Aldoory & Toth, 2004). There have been researching studies in the past prescribing the best practices of online communication, the power of engaging online with the customers with examples of few CEOs from western countries tweeting and using the social media as an extension of their workplace. But, there is less literature on the practices followed by C-Suite employees from developing countries like India and there is a lot of scope for research on how exactly the top level executives communicate. In next section, the adopted methodologies for this study are discussed.

METHODOLOGY

For this paper, Twitter data related to Ms. Kiran Mazumdar Shaw (CEO and Founder, Biocon) were collected between March 1 and April 25, 2014. During the period, R, an open source tool, was used to extract tweets from twitter through streaming API containing search words. The study involved analysis of three datasets namely Ms. Shaw's tweets, her tweets retweeted by other users and mentions for Ms. Shaw on Twitter. Data analyses were carried in three different phases to identify her communication style and discussion topics in tweet data. For that, few particular methods were used: word frequency analysis, metadiscourse, and content analysis. For that, we use another tool called NVivo which is a qualitative data analysis (QDA) computer software package. Results are presented and discussed in next section.

RESULTS AND FINDING

Content Analysis

Word Frequency Analysis

Word Frequency analysis is a method to quantitatively present how often words, phrases, and events occur. In the literature, researchers have used frequency analysis with words to examine significant meanings and characteristics (Ghiassi et al., 2013). Word Clouds are a visual representation of text data and help us to quickly understand the most prominent terms of discussion. It takes into account frequency of occurrence of the various terms and accordingly certain terms appear bolder with respect to the other terms in word clouds. For this paper, we demonstrated word clouds for three datasets as Ms. Shaw's tweets, her tweets retweeted by other users and mentions for Ms. Shaw on twitter (Figure 1), (Figure 2), (Figure 3).

These figures provide a comparison between Ms. Kiran Shaw's tweets and her tweets retweeted by other users. The prominent terms from Figure 1 indicate Ms. Kiran Shaw's social media communication is not limited to Biocon or pharma sector. Her tweets show continuous focus on emerging political landscape with aap, missionnamopm, political, parties coming out boldly in the word cloud. The individuals' mdpai, arunmsk and organizations bpac, timesofindia, bbmp are finding the equal reference. She tends to keep her discussion focused on multiple issues and tweets on length about them as we can see number of bolder terms emerging from Figure 1 compared to Figure 2. The conversation tends to be longer, probably involving two-way exchange of ideas which can be considered as the reason for the

Figure 1. Word cloud tweets posted by Ms. Kiran Shaw

India	59
Bangalore	58
Good	55
Thanks	51
Govt	47
Mdpai	39
Political	36
Time	36
Women	34
Yes	34
Support	32
Aap	29
Agree	29
Bioconlimited	28
Karnataka	26
Parties	26
missionnamopm	25
Indian	24
Change	23
Great	23

Figure 2. Word cloud for retweets posts shared by other users

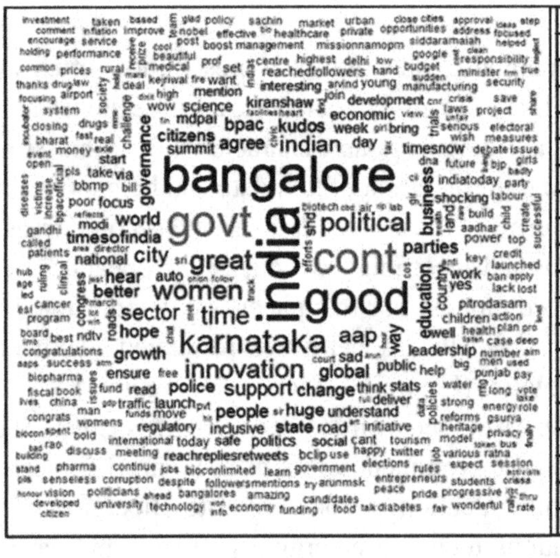

india	95
bangalore	84
good	74
cont	68
govt	67
karnataka	49
great	38
women	38
political	37
innovation	36
time	35
aap	29
city	29
indian	29
sector	26
support	25
global	23
agree	21
change	21
Kudos	21

discussion themes coming out so clearly in Figure 1. Compare this with Figure 2, where there are fewer prominent themes emerging out of the word cloud, and can be broadly identified as India, Bangalore, Karnataka, Political, Women, Innovation, Support. The themes which can be easily identified in both the datasets range from Bangalore, India, Govt, Karnataka, Women.

Figure 3. Word cloud for mentions by other users

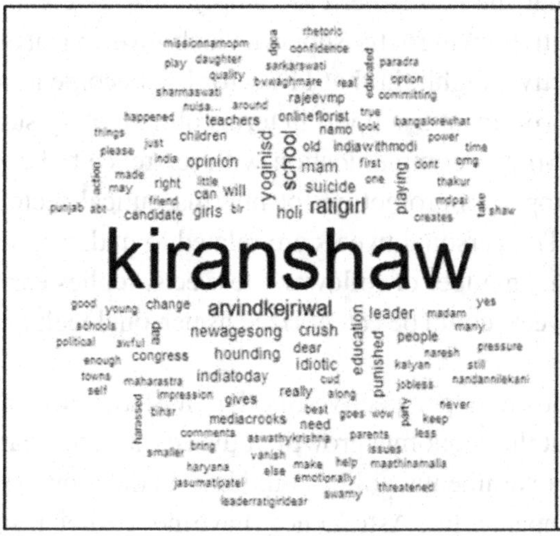

kiranshaw	149
arvindkejriwal	27
school	25
holi	14
suicide	14
yoginisd	13
girls	12
hounding	12
idiotic	12
leader	12
mam	12
opinion	12
indiatoday	11
playing	11
punished	11

The retweeted tweets by a user illustrate his interest or support for the messages shared by Ms. Kiran Shaw (Figure 2). With limited knowledge about the discussion topics from the word clouds, we can still assume maximum retweets are being done for issues of social concern - women, about communities – whether Bangalore, Karnataka, India. Also, the tweets addressed towards govt. are being retweeted to a large extent by her followers. At, the same time there are lesser retweets relating to her company Biocon or pharma sector by followers of Ms. Kiran Shaw. Her followers are retweeting about politics or say different political parties, which can be ascribed to the current scenario with general elections scheduled in the months of April-May'14. Also, dependent on the time frame for which analysis is being on Ms. Shaw's tweets, certain aspects would be highlighted more in the word cloud which might not be the case otherwise. To illustrate, the current period of discussion involved a lot of political activity on account of national elections and hence her communication involves reference to the same. In the subsequent portions, an effort has been put to study tweet contents manually in order to gain a better understanding of the themes of discussion. The tweets have been bifurcated on the basis of content to learn about the broader fields of her twitter communication.

Semantic Analysis

The semantic analysis provides the better understanding of the communication style of Ms. Kiran Shaw on Twitter. We divided tweets into three broad categories to understand tweet contents: Business (Table 1), Personal (Table 2), and Social (Table 3). The analysis demonstrates that she makes her case amply clear by comparing better research carried out in countries owing to the support which government offers. Findings, from a CEO's perspective, highlight that IT sector has become a model for development owing to the governmental support in terms of regulatory support offered over the last few years. So, her communication with reference to business includes the news related to company's promotion for pharmaceutical sector and views on governmental policy. The personal tweets are related to updating status, sharing photographs, increase in a number of followers, retweets, replies etc. over a period of time. The personal tweets could be categorized as personal feelings and status updates.

The findings show that for business perspective Ms. Kiran Shaw refers not to confine herself and is vocal about the economic growth of the country, the learning from the other sectors and how governmental policy could impact the growth story. In few tweets, she quotes how companies like AstraZeneca have moved their research facilities outside India owing to the poor research support which according to her

Table 1. Business category

	Promote/share company news	Promote/ share news for Pharmaceutical sector	Tweets/Retweets about other sectors	Comments on Governmental Policy/Regulations, Lobbying
Tweets	22	41	40	60
Re tweeted tweets	19	55	42	58

Table 2. Personal category

	Personal interests/feelings	Promote yourself/company	Status Update
Tweets	212	55	46
Retweeted tweets	264	38	50

Table 3. Social category

	Politics/Governance	Healthcare	Education	Women Empowerment	Other issues
Tweets	28	45	15	24	36
Re tweeted tweets	29	23	18	23	52

thwarts innovation and has a direct bearing on the economic growth of the country. Various topics in her tweets are presented in Figure 4. The Figure shows a wide range of topics on different categories discussed above. Her major discussion is about political activities by political parties such as 'arvind kejriwal vanish along congress', 'idiotic hounding related to namo', and 'political parties' rally and agendas'. She also discussed some shocking and sad news for a specific government (e.g. Karnataka government). For more information at that point, she was looking for more stats and hoped for good news and engaging news for better governance. In general elections, a transparent electoral process is much required for the common man.

Apart from political agendas, she felt a need of more research for criminal cases against women. Based on her discussion, it influences the growth of socio-economic development. Ms. Shaw also shared her views on equal opportunities for women. In the business category, she talked about jobs created by real states and start-ups. In her view, these sectors are changing things tremendously. Her posted content also showed her views on e-health and agri-biotechnology. In tweets, she showed her gratitude for different foundations for their support. Her discussion also raised the current society related issues, for instance, there was a suicidal case of the young girl in Bangalore. Another case was about few students being punished by the local school for playing Holi in Bangalore. In this way, Ms. Shaw talked about the safety issue in the city as a primary concern for society. Adding to it, she also shared her views on equality rights of education, especially in private sector. In summary, the semantic analysis highlights the variety of concerns and views of a woman CEO in a public platform. In next section, sentiment analysis has been carried to get a brief view of different emotions in her content.

Sentiment Analysis

Sentiment analysis shows the various emotions which were classified as anger, anticipation, disgust, fear, and joy. In this analysis, there is a score associated with each emotion based on the word sentiments (Figure 5). This combined score helps in identifying the major emotion which reflects the emotional personality of the person. The Figure shows that there are more positive emotions than negative which means that Ms. Shaw shares and discusses better happening content. People have more trust on her content in terms of views and opinions. On the other hand, there are emotions such as anger, disgust and sadness indicate her negative thoughts on various social issues, political agendas, and government policies. Such negative emotions lead to fear emotion highlighting her concern for growth and development. However, she is expecting few good moves from the government for a society which is reflected by anticipation emotion. In summary, she had positive thoughts about business related discussion and negative emotions with social issues.

Figure 4. Semantic Analysis

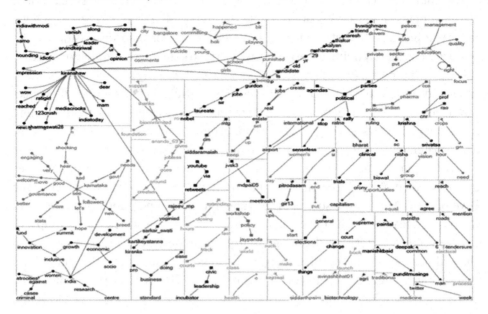

Figure 5. Emotion classification

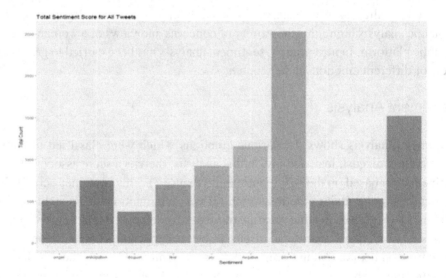

Metadiscourse Analysis

It is important to analyze the tweets irrespective of propositional content. Using metadiscourse, an attempt is made to understand the textual and interpersonal components by discerning the tweet contents. The approach, thereby, seeks to understand following appeals in the tweets made by Kiran Shaw: Rationality, Credibility, and Emotionality. These components together form pillars of effective communication stated by Aristotle. The textual discourse analysis of Ms. Kiran Shaw's tweets takes into account logical connectives, sequencers, frame makers and code glosses. The medium of communication reduces the scope of usage of endophoric markers (e.g. noted above, see below) and they have a minimal existence in the tweet sample. Therefore, the study disregards this category while understanding textual metadiscourse. The language of discourse on social media tends to be informal; it has been kept in mind to include all possible text tokens required for a complete analysis. The need to include (n, nd, &, and) while looking out for logical connective and in the tweet contents is on account of the informality of the communication medium.

A sample of 300 tweets was analyzed for Ms. Kiran Shaw to understand her communicative appeal. The model is highly adaptive since it makes use of the elementary components of effective communication, irrespective of the communication medium. The analysis was done on 4735 words in the 300 tweets sample. The research doesn't ignore the possibility of multiple categorizations of text tokens. However, in the current study, the primary purpose of the text token has been taken into account and double counting (considering them in multiple categories) has been avoided. Various categories of textual metadiscourse and text tokens corresponding to them are (shown in Table 4): logical connectives (and, or, but, however), sequencers (first, then, next, finally, number used for listing), frame makers (well, now, as a result, conclude), and code glosses (such as, namely, example, parentheses, for instance).

There is an attempt to generate a coherent and convincing piece of information, which is done with the help of above-mentioned text tokens. The quantitative analysis of these tokens helps us to understand the rational appeal of the tweets. The use of these tokens can determine what message Kiran Shaw wants to convey to the readers, influence them and restrict other possible understanding of the tweets. We also tried to compare the same with others CEOs speeches available in the literature (Table 5).

Table 4. Textual metadiscouse categorization frequency

	n	Per 100 words	% of total
Textual	109	2.30	36.82
Logical Connectives	81	1.71	27.36
Sequencers	8	0.17	2.70
Frame Markers	5	0.11	1.69
Code Glosses	15	0.32	5.07

Table 5. Comparison of textual metadiscourse between Ms. Shaw's tweets and other CEOs speeches

	Per 100 words	
	Kiran Shaw's tweets	CEO speeches
Textual	2.30	1.29
Logical Connectives	1.71	0.91
Sequencers	0.17	0.19
Frame Markers	0.11	0.11
Code Glosses	0.32	0.02

The finding shows, the textual metadiscourse is about 75% more in Ms. Shaw's tweets. This can be attributed to higher usage of logical connectives in her tweets. A close analysis of the tweets would illustrate her social media communication style as discussing multiple issues in her tweets. She prefers considering various possibilities/factors even while commenting/discussing issues and contributes towards a thoughtful discussion. In addition, she focuses on citing examples and highlighting certain important text in her tweets (read code glosses). The preferred use (0.32 vs. 0.2) indicates her propensity to use citations to convey her message, which also helps readers to easily understand and relate to her message. Further, the interpersonal discourse tends to evaluate the tenor of tweets such as the degree of reader involvement, commitment to shared information, writer's attitude etc. (Table 6). The interpersonal discourse analysis helps to understand the credibility and emotional appeal of tweets.

Ms. Kiran Shaw prefers to be considered as part of the general public, though throughout the discussion she dons the mantle of a corporate citizen, raises multiple issues of social issues in her tweets. The analysis shows that effective (or emotional) appeal tends to dominate the discourse of Ms. Kiran Shaw's communication on Twitter.

Table 6. Interpersonal discourse

	n	Per 100 words	% of total
Interpersonal	187	3.95	63.18
Hedges	2	0.04	0.68
Emphatics	6	0.13	2.03
Attributors	80	1.69	27.03
Attitude markers	12	0.25	4.05
Relational markers	87	1.84	29.39

In summary, Ms. Kiran Shaw's tweets can bring the spotlight to existing evils which can have a positive effect to bring about the governmental action. Her views can bring about systematic changes in the way society as a whole shares views on certain issues of public importance. The issues can vary in terms of significance and might have limited scope. In few of the tweets, she reflects strong support for the governmental setup and individuals for their visionary leadership shown at various levels. Ms. Kiran Shaw tweets are comparatively more about personal content (sharing information/feelings) than being related to business or social issues.

CONCLUSION

This study shows the various different perspective of discussion and opinions of an Indian woman CEO in her communication on Twitter. Interestingly, CEOs are more likely to share the newspaper article than the long blog post. As findings show the presence of broad different categories in Ms. Shaw's tweets, it reflects the balanced approach to creating content and engaging with public on social media. Her content is quite vocal about the social issues affecting the local community and even at the national level. She follows the accountability of the public relations professional to connect with different stakeholders (public and media) in relationship building activities. Her views also reflect lobbying for better support in research and development. This broader interaction strategy by a CEO may work better if it accompanies a suitable and strategic social media communication way involving severe ethical issues, and moral challenges.

LIMITATION AND FUTURE SCOPE

The current study analyzes an Indian CEOs twitter communication over a small period of time and takes into account only limited data. As the datasets are very limited due to restrictions imposed by Twitter APIs, it requires few advanced techniques to extract a larger set of tweets to generate more generalize results. It is proposed that further studies should move beyond content analysis or in-depth understandings such as influencer identification through network analysis.

REFERENCES

Aldoory, L., & Toth, E. (2004). Leadership and gender in public relations: Perceived effectiveness of transformational and transactional leadership styles. *Journal of Public Relations Research*, *16*(2), 157–183. doi:10.12071532754xjprr1602_2

Alghawi, I. A., Yan, J., & Wei, C. (2014). Professional or interactive: CEOs' image strategies in the microblogging context. *Computers in Human Behavior*, *41*, 184–189. doi:10.1016/j.chb.2014.09.027

Borghesi, R., Houston, J. F., & Naranjo, A. (2014). Corporate socially responsible investments: CEO altruism, reputation, and shareholder interests. *Journal of Corporate Finance*, *26*, 164–181. doi:10.1016/j.jcorpfin.2014.03.008

BRANDfog Survey. (2014). *The Global, Social CEO*. Retrieved April 20, 2014, from http://brandfog.com/CEOSocialMediaSurvey/BRANDfog_2014_CEO_Survey.pdf

Di Giuli, A., & Kostovetsky, L. (2014). Are red or blue companies more likely to go green? Politics and corporate social responsibility. *Journal of Financial Economics*, *111*(1), 158–180. doi:10.1016/j.jfineco.2013.10.002

Getting social: Social media in business. (n.d.). PwC. Retrieved March 19, 2014, from http://www.pwc.com/my/en/issues/socialmedia.html

Ghiassi, M., Skinner, J., & Zimbra, D. (2013). Twitter brand sentiment analysis: A hybrid system using n-gram analysis and dynamic artificial neural network. *Expert Systems with Applications*, *40*(16), 6266–6282. doi:10.1016/j.eswa.2013.05.057

Hutton, J. G., Goodman, M. B., Alexander, J. B., & Genest, C. M. (2001). Reputation management: The new face of corporate public relations? *Public Relations Review*, *27*(3), 247–261. doi:10.1016/S0363-8111(01)00085-6

Hwang, S. (2012). The strategic use of Twitter to manage personal public relations. *Public Relations Review*, *38*(1), 159–161. doi:10.1016/j.pubrev.2011.12.004

Hyland, K. (1998). Exploring corporate rhetoric: Metadiscourse in the CEO's letter. *Journal of Business Communication*, *35*(2), 224–244. doi:10.1177/002194369803500203

Jameson, D. A. (2014). Crossing Public-Private and Personal-Professional Boundaries How Changes in Technology May Affect CEOs' Communication. *Business and Professional Communication Quarterly*, *77*(1), 7–30. doi:10.1177/2329490613517133

Karaduman, İ. (2013). The effect of social media on personal branding efforts of top level executives. *Procedia: Social and Behavioral Sciences*, *99*, 465–473. doi:10.1016/j.sbspro.2013.10.515

Kim, J. N., & Rhee, Y. (2011). Strategic thinking about employee communication behavior (ECB) in public relations: Testing the models of megaphoning and scouting effects in Korea. *Journal of Public Relations Research*, *23*(3), 243–268. doi:10.1080/1062726X.2011.582204

Kim, S., Zhang, X. A., & Zhang, B. W. (2016). Self-mocking crisis strategy on social media: Focusing on Alibaba chairman Jack Ma in China. *Public Relations Review*, *42*(5), 903–912. doi:10.1016/j.pubrev.2016.10.004

Men, L. R. (2014). Why leadership matters to internal communication: Linking transformational leadership, symmetrical communication, and employee outcomes. *Journal of Public Relations Research*, *26*(3), 256–279. doi:10.1080/1062726X.2014.908719

Men, L. R. (2015). The internal communication role of the chief executive officer: Communication channels, style, and effectiveness. *Public Relations Review*, *41*(4), 461–471. doi:10.1016/j.pubrev.2015.06.021

Men, L. R., & Stacks, D. (2014). The effects of authentic leadership on strategic internal communication and employee-organization relationships. *Journal of Public Relations Research*, *26*(4), 301–324. doi:10.1080/1062726X.2014.908720

Men, L. R., & Tsai, W. H. S. (2016). Public engagement with CEOs on social media: Motivations and relational outcomes. *Public Relations Review*, *42*(5), 932–942. doi:10.1016/j.pubrev.2016.08.001

Park, J., Kim, H., Cha, M., & Jeong, J. (2011). *Ceo's apology in twitter: A case study of the fake beef labeling incident by e-mart*. Springer Berlin Heidelberg.

Porter, M. C., Anderson, B., & Nhotsavang, M. (2015). Anti-social media: Executive Twitter "engagement" and attitudes about media credibility. *Journal of Communication Management*, *19*(3), 270–287. doi:10.1108/JCOM-07-2014-0041

Rathore, A. K., & Ilavarasan, P. V. (2018). Social Media and Business Practices. In Encyclopedia of Information Science and Technology, Fourth Edition (pp. 7126-7139). IGI Global. doi:10.4018/978-1-5225-2255-3.ch619

Rathore, A. K., Ilavarasan, P. V., & Dwivedi, Y. K. (2016). Social media content and product co-creation: An emerging paradigm. *Journal of Enterprise Information Management*, 29(1), 7–18. doi:10.1108/JEIM-06-2015-0047

Report - 2013 Social CEO Report: Are America's Top CEOs Getting More Social? (n.d.). Domo. Retrieved March 3, 2014, from https://www.domo.com/learn/2013-social-ceo-report-are-americas-top-ceos-getting-more-social

Resick, C. J., Whitman, D. S., Weingarden, S. M., & Hiller, N. J. (2009). The bright-side and the dark-side of CEO personality: Examining core self-evaluations, narcissism, transformational leadership, and strategic influence. *The Journal of Applied Psychology*, 94(6), 1365–1381. doi:10.1037/a0016238 PMID:19916649

Rybalko, S., & Seltzer, T. (2010). Dialogic communication in 140 characters or less: How Fortune 500 companies engage stakeholders using Twitter. *Public Relations Review*, 36(4), 336–341. doi:10.1016/j.pubrev.2010.08.004

Schultz, F., Utz, S., & Göritz, A. (2011). Is the medium the message? Perceptions of and reactions to crisis communication via twitter, blogs and traditional media. *Public Relations Review*, 37(1), 20–27. doi:10.1016/j.pubrev.2010.12.001

Steyn, B. (2004). From strategy to corporate communication strategy: A conceptualisation. *Journal of Communication Management*, 8(2), 168–183. doi:10.1108/13632540410807637

Vidgen, R., Mark Sims, J., & Powell, P. (2013). Do CEO bloggers build community? *Journal of Communication Management*, 17(4), 364–385. doi:10.1108/JCOM-08-2012-0068

Waters, R. D., Burnett, E., Lamm, A., & Lucas, J. (2009). Engaging stakeholders through social networking: How nonprofit organizations are using Facebook. *Public Relations Review*, 35(2), 102–106. doi:10.1016/j.pubrev.2009.01.006

Watkins, B. A. (2017). Experimenting with dialogue on Twitter: An examination of the influence of the dialogic principles on engagement, interaction, and attitude. *Public Relations Review*, 43(1), 163–171. doi:10.1016/j.pubrev.2016.07.002

Weber Shandwick. (2012). *The social CEO: executives tell all*. Retrieved from. http://www.webershandwick.com/uploads/news/files/Social-CEO-Study.pdf

Compilation of References

2K Boston & 2K Australia. (2007). *BioShock* [Software]. USA: 2K Games.

AADC. (2013). *About us*. Retrieved November 14, 2014, from http://www.AADC.ae/en/about-us.aspx

Aarseth, E. (2012). A narrative theory of games. In M.S El-Nasr, M. Consalvo, & S. Feiner (Eds.), *Proceedings of the International Conference on the Foundations of Digital Games* (pp. 129–133). New York, NY: ACM.

Abdul Razak, S.B., & Latip, N.A.B. (2016)."Factors That Influence The Usage of Social Media In Marketing. *Journal of Research in Business and Management*, *4*(2), 1-7.

Abel, F., Henze, N., Herder, E., & Krause, D. (2010). Interweaving public user profiles on the web. *User Modeling, Adaptation, and Personalization*, 16-27.

Abel, F., Araújo, S., Gao, Q., & Houben, G. J. (2011, June). Analyzing cross-system user modeling on the social web. In *International Conference on Web Engineering* (pp. 28-43). Springer. 10.1007/978-3-642-22233-7_3

Adamic, L., Buyukkokten, O., & Adar, E. (2003). A social network caught in the web. *First Monday*, *8*(6). doi:10.5210/fm.v8i6.1057

Adams, E. W. (2013). *Resolutions to Some Problems in Interactive Storytelling* (Doctoral dissertation). University of Teesside, Middlesbrough, UK.

Adams, R. J., & Khoo, S.-T. (1996). *QUEST:The Interactive Test Analysis System* (Vol. 1). Melbourne: Australian Council for Educational Research.

Adomavicius, G., & Tuzhilin, A. (2005). Toward the next generation of recommender systems: A survey of the state-of-the-art and possible extensions. *IEEE Transactions on Knowledge and Data Engineering*, *17*(6), 734–749. doi:10.1109/TKDE.2005.99

Agrawal, J., & Kamakura, W. A. (1995). The Electronic Worth of Celebrity Endorsers: An Event Study Analysis. *Journal of Marketing*, *59*(July), 56–62. doi:10.2307/1252119

Ahluwalia, R., Robert, B., & Unnava, H. (2000). Consumer Response to Negative Publicity: The Moderating Role of Commitment. *JMR, Journal of Marketing Research, 37*(2), 203–214. doi:10.1509/jmkr.37.2.203.18734

Ahmed, A. A. A. (2010). Online privacy concerns among social networks' users. *Cross-Cultural Communication, 6,* 74–89.

Ahmed, A. A. A. (2015). "Sharing is Caring": Online Self-Disclosure, Offline Social Support, and SNS in the UAE. *Contemporary Review of Middle East, 2*(3), 192–219. doi:10.1177/2347798915601574

Ahn, J. (2012). Teenagers' Experiences with Social Network Sites: Relationships to Bridging and Bonding Social Capital. *The Information Society, 28*(2), 99–109. doi:10.1080/01972243.2011.649394

Ahn, Y. Y., Han, S., Kwak, H., Moon, S., & Jeong, H. (2007, May). Analysis of topological characteristics of huge online social networking services. In *Proceedings of the 16th international conference on World Wide Web* (pp. 835-844). ACM. 10.1145/1242572.1242685

Aiello, L. M., Barrat, A., Cattuto, C., Ruffo, G., & Schifanella, R. (2010, August). Link creation and profile alignment in the aNobii social network. In *Social Computing (SocialCom), 2010 IEEE Second International Conference on* (pp. 249-256). IEEE. 10.1109/SocialCom.2010.42

Al Jenaibi, B. N. A. (2011). Use of Social Media in the United Arab Emirates: An Initial Study. *Global Media Journal Arabian Edition, 1*(2), 3-27. Retrieved from: http://www.gmjme.com/gmj_custom_files/volume1_issue2/articles_in_english/volume1-issue2-article-3-27.pdf

Aldoory, L., & Toth, E. (2004). Leadership and gender in public relations: Perceived effectiveness of transformational and transactional leadership styles. *Journal of Public Relations Research, 16*(2), 157–183. doi:10.12071532754xjprr1602_2

Alghawi, I. A., Yan, J., & Wei, C. (2014). Professional or interactive: CEOs' image strategies in the microblogging context. *Computers in Human Behavior, 41,* 184–189. doi:10.1016/j.chb.2014.09.027

Alim, S., Abdulrahman, R., Neagu, D., & Ridley, M. (2011). Online social network profile data extraction for vulnerability analysis. *International Journal of Internet Technology and Secured Transactions, 3*(2), 194–209. doi:10.1504/IJITST.2011.039778

Al-Jenaibi, B. (2011). The Use of Social Media in the United Arab Emirates – An Initial Study. *European Journal of Soil Science, 23*(1), 84–100.

Allen, M. W. (2016). *Michael Allen's Guide to e-Learning: Building Interactive, Fun, and Effective Learning Programs for Any Company.* John Wiley & Sons. doi:10.1002/9781119176268

Allwood, J., Traum, D., & Jokinen, K. (2000). Cooperation, Dialogue and Ethics. *International Journal of Human-Computer Studies, 53*(6), 871–914. doi:10.1006/ijhc.2000.0425

Alreyaee, S.I., & Ahmed, A. (2015). Trends in Social Media Usage: An Investigation of its Growth in the Arab World. *International Journal of Virtual Communities and Social Networking Archive, 7*(4), 45-56.

Alwi, A., & McKay, E. Investigating an online museum's information system: Instructional design for effective human-computer interaction. In P. I. Demetrios, G. Sampson, & J. M. Spector (Eds.), *D.Ifenthaler, Kinshuk*. Germany: Multiple Perspectives on Problem Solving and Learning in the Digital Age.

Angeletou, S. (2006). Information Systems for e-Government - A Case in the Hellenic Ministry of Education. *International Journal of Public Information Systems, 2*(2), 55–68.

Asperen, M., Rooij, P., & Dijkmans, C. (2017). Engagement-based loyalty: The effects of social media engagement on customer loyalty in the travel industry. *International Journal of Hospitality & Tourism Administration, 17*(4), 1–17.

Asur, S., & Huberman, B. A. (2010). *Predicting the Future With Social Media*. Retrieved March 22, 2014, from http://www.hpl.hp.com/research/scl/papers/socialmedia/socialmedia.pdf

Aubrey, S., Chattopadhyay, S., & Rill, L. (2008). *Are Facebook Friends Like Face-to-Face Friends: Investigating Relations Between the Use of Social Networking Websites and Social capital*. Paper presented at the annual meeting of the International Communication Association.

Aubrey, F. S., & Rill, L. (2013). Investigating Relations between Facebook Use and Social capital Among College Undergraduates. *Communication Quarterly, 61*(4), 479–496. doi:10.1080/01463373.2013.801869

Aylett, R., & Louchart, S. (2007). Being there: Participants and spectators in interactive narrative. In M. Cavazza & S. Donikian (Eds.), *Virtual Storytelling. Using Virtual Reality Technologies for Storytelling. Proceedings of the Fourth International Conference (ICVS2007)* (Vol. 4871, pp. 117–128). Berlin, Germany: Springer-Verlag. 10.1007/978-3-540-77039-8_10

Aylett, R., Louchart, S., & Weallans, A. (2011). Research in interactive drama environments, role-play and story-telling. In M. Si, D. Thue, E. André, J. Lester, J. Tanenbaum, & V. Zammitto (Eds.), *Interactive Storytelling: 4th International Conference on Interactive Digital Storytelling (ICIDS 2011)* (Vol. 7069, pp. 1–12). Berlin, Germany: Springer-Verlag. 10.1007/978-3-642-25289-1_1

Ayyad, K. (2011). Internet Usage vs. Traditional Media Usage among University Students in the United Arab Emirates. *Journal of Arab & Muslim Media Research, 4*(1), 41–61. doi:10.1386/jammr.4.1.41_1

Bagley, C. A., & Heltne, M. (2003). *Cooperative Teams*. Paper presented at the Collaboration Conference, Radisson South, Bloomington, MN.

Bakkar, M. N. (2016). *An Investigation of Mobile Healthcare (mHealcare) Training Design for Healthcare Empoyees in Jordan*. RMIT University, School of Business Information Technology and Logistics.

Banczyk, B., & Senokozlieva, M. (2008). *The "Wurst" Meets "Fatless" in MySpace: The Relationship Between Self-Esteem, Personality, and Self-Presentation in an Online Community.* International Communication Association, Annual Meeting.

Barefah, A., & McKay, E. (2017). Evaluating the effectiveness of teaching information systems courses: A Rasch measurement approach. *Proceedings of the 9th Annual International Scientific Conference (DLCC).*

Barefah, A., & McKay, E. (2018). A Rasch-driven validation for measuring the effectiveness of eLearning Applications in Higher Education. *Seventh International Conference on Probabilistic Models for Measurement.*

Bargh, J., & McKenna, K. (2004). The Internet and Social Life. *Annual Review of Psychology, 55*(1), 573–590. doi:10.1146/annurev.psych.55.090902.141922 PMID:14744227

Barhemmati, N., & Ahmad, A. (2015). Effects of social network marketing (SNM) on consumer purchase behavior through customer engagement. *Journal of Advanced Management Science, 3*(4), 307–311. doi:10.12720/joams.3.4.307-311

Bell, J. (2010). *Doing your research project – A guide for first-time researchers.* Buckingham, UK: OUP.

Benevenuto, F., Rodrigues, T., Cha, M., & Almeida, V. (2009, November). Characterizing user behavior in online social networks. In *Proceedings of the 9th ACM SIGCOMM conference on Internet measurement conference* (pp. 49-62). ACM. 10.1145/1644893.1644900

Berger, J., Sorensen, A. T., & Rasmussen, S. J. (2010). Positive Effects of Negative Publicity: When Negative Reviews Increase Sales. *Marketing Science, 29*(5), 815–827. doi:10.1287/mksc.1090.0557

Berkovsky, S., Kuflik, T., & Ricci, F. (2008). Mediation of user models for enhanced personalization in recommender systems. *User Modeling and User-Adapted Interaction, 18*(3), 245–286. doi:10.100711257-007-9042-9

Biagi, S. (2004). *Media Impact.* Wadsworth Publications.

Blazhenkova, O., & Kozhevnikov, M. (2009). The New Object-Spatial-Verbal Cognitive Style Model: Theory and measurement. *Applied Cognitive Psychology, 23*(5), 638–663. doi:10.1002/acp.1473

Blizzard Entertainment. (2004). *World of Warcraft* [Software]. USA: Blizzard Entertainment.

Boase, J., Horrigan, J., Wellman, B., & Rainie, L. (2006). The Strength of Internet Ties: The internet and E-mail Aid Users in Maintaining Their Social Networks and Provide Pathways to Help When People Face Big Decisions. *Pew Internet and American Life Project.* Retrieved from: http://www.pewinternet.org/files/old-media/Files/Reports/2006/PIP_Internet_ties.pdf.pdf

Bond, T. G., & Fox, C. M. (2015). *Applying the Rasch Model: Fundamental measurement in the human services* (3rd ed.). New York: Routledge, Taylor & Francis.

Borghesi, R., Houston, J. F., & Naranjo, A. (2014). Corporate socially responsible investments: CEO altruism, reputation, and shareholder interests. *Journal of Corporate Finance, 26*, 164–181. doi:10.1016/j.jcorpfin.2014.03.008

Bouraga, S., Jureta, I., & Faulkner, S. (2015), Rules for relevant notification recommendations to online social networks users. In *ASE Eighth International Conference on Social Computing (SocialCom)*. Academic Press.

Bouraga, S., Jureta, I., & Faulkner, S. (2015). An empirical study of notifications' importance for online social network users. *Social Network Analysis and Mining, 5*(1), 51. doi:10.100713278-015-0293-x

Bowden, J. L.-H. (2009). The Process of Customer Engagement: A Conceptual Framework. *Journal of Marketing Theory and Practice, 17*(1), 63–74. doi:10.2753/MTP1069-6679170105

boyd, d. (2008). *Taken Out of Context: American Teen Sociality in Networked Publics* (PhD Dissertation). University of California-Berkeley, School of Information.

Boyd, D. M., & Ellison, N. B. (2007). Social Network Sites: Definition, History, and Scholarship. *Journal of Computer-Mediated Communication, 13*(1), 210–230. doi:10.1111/j.1083-6101.2007.00393.x

Branch, R. M., & Kopcha, T. J. (2014). *Instructional design models. In Handbook of research on educational communications and technology* (pp. 77–87). Springer.

BRANDfog Survey. (2014). *The Global, Social CEO*. Retrieved April 20, 2014, from http://brandfog.com/ CEOSocialMediaSurvey/BRANDfog_2014_CEO_Survey.pdf

Branson, R. K. (1975). *Interservice Procedures for Instructional Systems Development. Executive Summary and Model*. DTIC Document.

Brodie, R. J., Hollebeek, L. D., Jurić, B., & Ilić, A. (2011). Customer Engagement: Conceptual domain, fundamental preposition and implication of research. *Journal of Service Research, 14*(3), 252–271. doi:10.1177/1094670511411703

Burke, M., Kraut, R., & Marlow, C. (2011). *Social Capital on Facebook: Differentiating Uses and Users*. CHI 2011, Vancouver, Canada.

Burke, M., Marlow, C., & Lento, T. (2010). Social Network Activity and Social Well-Being. In *Proceeding of the 2010 ACM Conference on Human Factors in Computing Systems*. New York, NY: ACM.

Castells, M. (2009). *Rise of the Network Society*. Blackwell Publishers. doi:10.1002/9781444319514

Caverlee, J., & Webb, S. (2008, March). *A Large-Scale Study of MySpace: Observations and Implications for Online Social Networks*. ICWSM.

Centola, D. (2010). The spread of behavior in an online social network experiment. *Science, 329*(5996), 1194-1197.

Chao, J. T., Parker, K. R., & Fontana, A. (2011). Developing an interactive social media based learning environment. *Issues in Informing Science and Information Technology*, *8*, 323–334. doi:10.28945/1421

Chatman, S. (1978). *Story and Discourse: Narrative Structures in Fiction and Film*. Ithaca, NY: Cornell University Press.

Chen, J., Kiremire, A. R., Brust, M. R., & Phoha, V. V. (2014). Modeling online social network users' profile attribute disclosure behavior from a game theoretic perspective. *Computer Communications*, *49*, 18–32. doi:10.1016/j.comcom.2014.05.001

Chen, J., Xu, H., & Whinston, A. B. (2011a). Moderated online communities and quality of user-generated content. *Journal of Management Information Systems*, *28*(2), 237–268. doi:10.2753/MIS0742-1222280209

Chen, Y., Fay, S., & Wang, Q. (2011b). The role of marketing in social media: How online consumer reviews evolve. *Journal of Interactive Marketing*, *25*(2), 85–94. doi:10.1016/j.intmar.2011.01.003

Cheong, H. J., & Morrison, M. A. (2008). Consumers' reliance on product information and recommendations found in UGC". *Journal of Interactive Advertising*, *8*(2), 38–49. doi:10.1080/15252019.2008.10722141

Cheung, C. M. K., Lee, M. K. O., & Rabjohn, N. (2008). The impact of electronic word of-mouth: The adoption of online opinions in online customer communities. *Internet Research*, *18*(3), 229–247. doi:10.1108/10662240810883290

Chevalier, J. A., & Mayzlin, D. (2006). The Effect of Word of Mouth on Sales: Online Book Reviews. *JMR, Journal of Marketing Research*, *43*(August), 345–354. doi:10.1509/jmkr.43.3.345

Chi, H.-H. (2011). Interactive Digital Advertising VS. Virtual Brand Community: Exploratory Study of User Motivation and Social Media Marketing Responses in Taiwan. *Journal of Interactive Advertising*, *12*(1), 44–61. doi:10.1080/15252019.2011.10722190

Choi, S. M., Kim, Y., Sung, Y., & Sohn, D. (2011). Bridging or Bonding? A cross-cultural study of social relationships in social networking sites. *Information Communication and Society*, *14*(1), 107–129. doi:10.1080/13691181003792624

Chua, S.-C., & Choib, S. M. (2010). Social capital and self-presentation on social networking sites: A comparative study of Chinese and American young generations. *Chinese Journal of Communication*, *3*(4), 402–420. doi:10.1080/17544750.2010.516575

Chu, S. C., & Kim, Y. (2011). Determinants of Consumer engagement in electronic word of mouth (eWOM) in social networking sites". *International Journal of Advertising*, *30*(1), 47–75. doi:10.2501/IJA-30-1-047-075

Clark, R. C., Nguyen, F., & Sweller, J. (2006). *Efficiency in Learning: Evidence-based Guidelines to Manage Cognitive Load*. San Francisco: Wiley & Sons.

Colás, P., & ... . (2013). Young People and Social Networks: Motivations and Preferred Use. *Scientific Journal of Media Education, 2*(40), 15–23.

Coleman, J. S. (1988). Social Capital in the Creation of Human Capital. *American Journal of Sociology, 94*, 95–121. doi:10.1086/228943

Comer, D. (2009). *The Internet Book*. Prentice Hall.

Cooper-Smith, L. B., & McKay, E. (2015). *A Comparative Analysis of Tools Used to Measure Engagement in Online Collaborative Learning Activities.* Paper presented at the 2015 IEEE 3rd International Conference on MOOCS, Innovation and Technology in Education (MTE), Amritsar, India.

Creswell, J. (2012). *Educational research: Planning, conducting, and evaluating quantitative and qualitative research*. Prentice Hall.

Croucher, S. M. (2011). Social Networking and Cultural Adaptation: A Theoretical Model. *Journal of International and Intercultural Communication, 4*(4), 259–264. doi:10.1080/17513 057.2011.598046

Dahi, M., & Ezziane, Z. (2015). Measuring e-government adoption in Abu Dhabi with Technology Acceptance Model (TAM). *International Journal of Electronic Governance, 7*(3), 206–231. doi:10.1504/IJEG.2015.071564

Daugherty, T., Eastin, M. S., & Bright, L. (2008). Exploring consumer motivations for creating user-generated content. *Journal of Interactive Advertising, 8*(2), 16–25. doi:10.1080/15252019 .2008.10722139

Dehghani, M., & Tumer, M. (2015). A research on effectiveness of Facebook advertising on enhancing purchase intention of consumers. *Computers in Human Behavior, 49*(1), 597–600. doi:10.1016/j.chb.2015.03.051

Dewan, S., & Ramaprasad, J. (2014). Social Media, Traditional Media, and Music Sales. *Management Information Systems Quarterly, 38*(1), 101–121. doi:10.25300/MISQ/2014/38.1.05

Di Giuli, A., & Kostovetsky, L. (2014). Are red or blue companies more likely to go green? Politics and corporate social responsibility. *Journal of Financial Economics, 111*(1), 158–180. doi:10.1016/j.jfineco.2013.10.002

Dickey, I. J., & Lewis, W. F. (2011). *An Overview of Digital Media and Advertising*. Information Science Reference.

Dick, W. O., Carey, L., & O'Carey, J. (2004). *The Systematic Design of Instruction* (6th ed.). Allyn & Bacon.

Donath, J., & Boyd, D. (2004). Public Displays of Connection. *BT Technology Journal, 22*(4), 71–82. doi:10.1023/B:BTTJ.0000047585.06264.cc

Ducheneaut, N., Yee, N., Nickell, E., & Moore, R. J. (2006, April). Alone together?: exploring the social dynamics of massively multiplayer online games. In *Proceedings of the SIGCHI conference on Human Factors in computing systems* (pp. 407-416). ACM. 10.1145/1124772.1124834

Dudenhoffer, C. (2012). Pin it!: Pinterest as a library marketing and information literacy tool. *College & Research Libraries News, 73*(6), 328–332. doi:10.5860/crln.73.6.8775

Dunbar, R. I., Arnaboldi, V., Conti, M., & Passarella, A. (2015). The structure of online social networks mirrors those in the offline world. *Social Networks, 43*, 39–47. doi:10.1016/j.socnet.2015.04.005

Dunne, A., Lawlor, M. A., & Rowley, J. (2010). Young People's Use of Online Social Networking Sites: A Uses and Gratifications Perspective. *Journal of Research in Interactive Marketing, 4*(1), 46–58. doi:10.1108/17505931011033551

Dwyer, C., Hiltz, S. R., & Passerini, K. (2007, August). Trust and Privacy Concern Within Social Networking Sites: A Comparison of Facebook and MySpace. In AMCIS (p. 339). Academic Press.

Dwyer, C., Hiltz, S., & Passerini, K. (2007). Trust and privacy concern within social networking sites: A comparison of Facebook and MySpace. *AMCIS 2007 Proceedings, 339*.

Dyer, P. (2013). The top benefits of social media marketing. *Social Media Today*. Retrieved February 6, 2016, from http://socialmediatoday.com/pamdyer/1568271/top-benefits-social-media-marketing-infographic

Edwards, S. M. (2011). A social media mindset. *Journal of Interactive Advertising, 12*(1), 1–3. doi:10.1080/15252019.2011.10722186

Ehrhardt, B. (Producer), Kragh-Jacobsen, S. (Director), Levring, K. (Director), Vinterberg, T. (Director), & von Trier, L. (Director). (1999, December 31). *D-dag* [Television broadcast]. Denmark: DR.

Elberse, A. (2007). The Power of Stars: Do Star Actors Drive the Success of Movies? *Journal of Marketing, 71*(October), 102–120. doi:10.1509/jmkg.71.4.102

Eldon, E. (2009). *2008 growth puts Facebook in better position to make money*. Retrieved February 6, 2016, from http://venturebeat.com/2008/12/18/2008-growth-puts-facebook-in-better-position-to-make-money/

Ellison, N. (2008). Introduction: Reshaping campus communication and community through social network sites. In The ECAR study of undergraduate students and information technology (vol. 8, pp. 19–32). Boulder, CO: Educause.

Ellison, N., Seinfield, C., & Lampe, C. (2006). *Spatially Bounded Online Social Networks and Social Capital: The Role of Facebook*. Paper presented at the Annual Conference of the International Communication Association (ICA), Dresden, Germany.

Ellison, N., Seinfield, C., & Lampe, C. (2007). The Benefits of Facebook "Friends": Social Capital and College Students' Use of Online Social Network Sites. *Journal of Computer-Mediated Communication, 12*(4), 1143–1168. doi:10.1111/j.1083-6101.2007.00367.x

Ellison, N., Seinfield, C., & Lampe, C. (2011). Connection Strategies: Social Capital Implications of Facebook-Enabled Communication Practices. *New Media & Society, 13*(6), 873–892. doi:10.1177/1461444810385389

Ellison, N., Vitak, J., Gray, R., & Lampe, C. (2014). Cultivating Social Resources on Social Network Sites: Facebook Relationship Maintenance Behaviors and Their Role in Social Capital Processes. *Journal of Computer-Mediated Communication, 19*(4), 855–870. doi:10.1111/jcc4.12078

Emanuel, L., Bevan, C., & Hodges, D. (2013, April). What does your profile really say about you?: privacy warning systems and self-disclosure in online social network spaces. In CHI'13 Extended Abstracts on Human Factors in Computing Systems (pp. 799-804). ACM.

Episode Interactive. (2013). *Episode* [Software]. USA: Episode Interactive. Available from https://home.episodeinteractive.com/

Ermi, L. & Mäyrä, F. (2005). Fundamental components of the gameplay experience: Analysing immersion. *Worlds in Play, 37*(2).

Ezziane, Z., & Al Kaabi, A.A. (2015). Impact of Social Media towards improving Productivity at AADC. *International Journal of Virtual Communities and Social Networking Archive, 7*(4), 1-22.

Ezziane, Z., & Al Shamisi, A. (2013). Improvement of the organizational performance through compliance with best practices in Abu-Dhabi. *International Journal of IT/Business Alignment and Governance, 4*(2), 19–36. doi:10.4018/ijitbag.2013070102

Facebook. (n.d.). Retrieved from www.facebook.com

Fire, M., Goldschmidt, R., & Elovici, Y. (2014). Online social networks: Threats and solutions. *IEEE Communications Surveys and Tutorials, 16*(4), 2019–2036. doi:10.1109/COMST.2014.2321628

Flanagin, A. J., Metzger, M. J., Pure, R., & Markov, A. (2011). User-generated ratings and the evaluation of credibility and product quality in ecommerce transactions. In *Proceedings of the 44th Hawaii International Conference of System Sciences*. ACM Digital Library. 10.1109/HICSS.2011.474

Forbes. (2017). *4Tips To Help Your Business Flourish On Social Media*. Retrieved from https://www.forbes.com/sites/jpmorganchase/ 2017/03/20/4-tips-to-help-your-business-flourish-on-socialmedia/#605da1987dd2

Franzblau, B. (Producer), & Bejan, B. (Director). (1992). *I'm Your Man* [Motion picture]. USA: Interfilm Technologies.

Fu, F., Liu, L., & Wang, L. (2008). Empirical analysis of online social networks in the age of Web 2.0. *Physica A, 387*(2), 675–684. doi:10.1016/j.physa.2007.10.006

Fukuyama, F. (2001). Social Capital, Civil Society and Development. *Third World Quarterly*, *22*(1), 7–20. doi:10.1080/713701144

Gadamer, H.-G. (2004). Truth and Method (2nd ed.). London, UK: Bloomsbury Academic.

Genette, G. (1980). *Narrative Discourse: An essay in method*. New York, NY: Cornell University Press.

Gentle, A. (2009). *Conversation and Community: The Social Web for documentation*. Fort Collins: XML Press.

Gershuny, J. (2002). Social Leisure and Home IT: A Panel Time-Diary Approach. *IT & Society*, *1*, 54–72.

Getting social: Social media in business. (n.d.). PwC. Retrieved March 19, 2014, from http://www.pwc.com/my/en/issues/socialmedia.html

Ghafoor, F., & Niazi, M. A. (2016). Using social network analysis of human aspects for online social network software: A design methodology. *Complex Adaptive Systems Modeling*, *4*(1), 14. doi:10.118640294-016-0024-9

Ghiassi, M., Skinner, J., & Zimbra, D. (2013). Twitter brand sentiment analysis: A hybrid system using n-gram analysis and dynamic artificial neural network. *Expert Systems with Applications*, *40*(16), 6266–6282. doi:10.1016/j.eswa.2013.05.057

Gilbert, E., Bakhshi, S., Chang, S., & Terveen, L. (2013, April). I need to try this?: a statistical overview of pinterest. In *Proceedings of the SIGCHI conference on human factors in computing systems*(pp. 2427-2436). ACM. 10.1145/2470654.2481336

Gill, P., Arlitt, M., Li, Z., & Mahanti, A. (2007, October). Youtube traffic characterization: a view from the edge. In *Proceedings of the 7th ACM SIGCOMM conference on Internet measurement* (pp. 15-28). ACM. 10.1145/1298306.1298310

Golbeck, J. (2009). Trust and nuanced profile similarity in online social networks. *ACM Transactions on the Web*, *3*(4), 12. doi:10.1145/1594173.1594174

Golder, S. A., Wilkinson, D. M., & Huberman, B. A. (2007). Rhythms of social interaction: Messaging within a massive online network. *Communities and technologies 2007*, 41-66.

Gopsill, J. A., McAlpine, H. C., & Hicks, B. J. (2013). A social media framework to support engineering design communication. *Advanced Engineering Informatics*, *27*(4), 580–597. doi:10.1016/j.aei.2013.07.002

Granovetter, M. (1973). The Strength of Weak Ties. *American Journal of Sociology*, *78*(6), 1360–1380. doi:10.1086/225469

Grant, C. (2007). *Uncertainty and communication*. Basingstoke, UK: Palgrave MacMillan. doi:10.1057/9780230222939

Greenhow, C., & Robelia, B. (2009). Old Communication, New Literacies: Social Network Sites as Social Learning Resources. *Journal of Computer-Mediated Communication, 14*(4), 1130–1161. doi:10.1111/j.1083-6101.2009.01484.x

Guha, S., Tang, K., & Francis, P. (2008, August). NOYB: Privacy in online social networks. In *Proceedings of the first workshop on Online social networks* (pp. 49-54). ACM. 10.1145/1397735.1397747

Guo, L., Tan, E., Chen, S., Zhang, X., & Zhao, Y. E. (2009, June). Analyzing patterns of user content generation in online social networks. In *Proceedings of the 15th ACM SIGKDD international conference on Knowledge discovery and data mining* (pp. 369-378). ACM. 10.1145/1557019.1557064

Gurses, S., Rizk, R., & Gunther, O. (2008). Privacy design in online social networks: Learning from privacy breaches and community feedback. *ICIS 2008 Proceedings*, 90.

Hair, J., Anderson, R., Tatham, & Black, W. (1998). Multivariate Data Analysis (5th ed.). PHI.

Hampton, K., & Wellman, B. (2003). Neighboring in Netville: How the Internet Supports Community and Social Capital in a Wired Suburb. *City & Community, 2*(4), 277–311. doi:10.1046/j.1535-6841.2003.00057.x

Hanafizadeh, P., Ravasanm, A. Z., Nabavi, A., & Mehrabioun, M. (2012). A Literature Review on the Business Impacts of Social Network Sites. *International Journal of Virtual Communities and Social Networking, 4*(1), 46–60. doi:10.4018/jvcsn.2012010104

Haythornthwaite, C. (2000). Online Personal Networks. *New Media & Society, 2*(2), 195–226. doi:10.1177/14614440022225779

Haythornthwaite, C. (2005). Social networks and Internet connectivity effects. *Information Communication and Society, 8*(2), 125–147. doi:10.1080/13691180500146185

Heckmann, D., Schwartz, T., Brandherm, B., Schmitz, M., & von Wilamowitz-Moellendorff, M. (2005). GUMO-the general user model ontology. Lecture Notes in Computer Science, 3538, 428.

Heer, J., & Boyd, D. (2005, October). Vizster: Visualizing online social networks. In *Information Visualization, 2005. INFOVIS 2005. IEEE Symposium on* (pp. 32-39). IEEE.

Herman, E. (2009). The Propaganda Model. *Journalism Studies, 1*(1).

Hofer, M., & Aubert, V. (2013). Perceived Bonding and Bridging Social Capital on Twitter: Differentiating between Followers and Followees. *Computers in Human Behavior, 29*(6), 2134–2142. doi:10.1016/j.chb.2013.04.038

Hutton, J. G., Goodman, M. B., Alexander, J. B., & Genest, C. M. (2001). Reputation management: The new face of corporate public relations? *Public Relations Review, 27*(3), 247–261. doi:10.1016/S0363-8111(01)00085-6

Hwang, S. (2012). The strategic use of Twitter to manage personal public relations. *Public Relations Review*, *38*(1), 159–161. doi:10.1016/j.pubrev.2011.12.004

Hyland, K. (1998). Exploring corporate rhetoric: Metadiscourse in the CEO's letter. *Journal of Business Communication*, *35*(2), 224–244. doi:10.1177/002194369803500203

India, Q. (2017). *It's official, Narendra Modi is the most followed world leader on Facebook.* Accessed on 22 -10- 2017, retrieved from https://qz.com/917170/its-official-narendra-modi-is-the-most-followed-world-leader-on-facebook/

Irrational Games. (2013). *BioShock Infinite* [Software]. USA: 2K Games.

Islam, K. M (2006). Social Capital and Health: Does Egalitarianism Matter?: A Literature Review. *International Journal for Equity in Health*, *5*(3), 1–28. PMID:16597324

Jakobson, R. (1960). Linguistics and poetics. In T. A. Sebeok (Ed.), *Style in Language* (pp. 350–377). Cambridge, MA: The MIT Press.

Jameson, D. A. (2014). Crossing Public-Private and Personal-Professional Boundaries How Changes in Technology May Affect CEOs' Communication. *Business and Professional Communication Quarterly*, *77*(1), 7–30. doi:10.1177/2329490613517133

Java, A., Song, X., Finin, T., & Tseng, B. (2007, August). Why we twitter: understanding microblogging usage and communities. In *Proceedings of the 9th WebKDD and 1st SNA-KDD 2007 workshop on Web mining and social network analysis* (pp. 56-65). ACM. 10.1145/1348549.1348556

Johnston, K. (2013). Social Capital: The Benefit of Facebook 'Friends'. *Información Tecnológica*, *32*(1), 24–36.

Jonas, J. R. O. (2010). Source credibility of company-produced and user generated content on the internet: An exploratory study on the Filipino youth. *Philippine Management Review*, *17*, 121–132.

Jones, E., Busch, P., & Dacin, P. (2003). Firm Market Orientation and Salesperson Customer Orientation: Interpersonal and Intrapersonal Influences on Customer Service and Retention in Business-to-Business Buyer-Seller Relationships. *Journal of Business Research*, *56*(4), 323–340. doi:10.1016/S0148-2963(02)00444-7

Kalas, L. (Producer), & Činčera, R. (Director). (1967). *Kinoautomat* [Motion picture]. Czechoslovakia.

Kaplan, A. M., & Haenlein, M. (2010). Users of the world, unite! The challenges and opportunities of Social Media. *Business Horizons*, *53*(1), 59–68. doi:10.1016/j.bushor.2009.09.003

Karaduman, İ. (2013). The effect of social media on personal branding efforts of top level executives. *Procedia: Social and Behavioral Sciences*, *99*, 465–473. doi:10.1016/j.sbspro.2013.10.515

Karampelas, P. (2012). *Techniques and tools for designing an online social network platform.* Springer.

Karuppasamy, G., Anwar, A., Bhartiya, A., Sajjad, S., Rashid, M., Mathew, E., ... Sreedharan, J. (2013). Use of Social Networking Sites among University Students in Ajman, United Arab Emirates. *Nepal Journal of Epidemiology.*, *3*(2), 245–250. doi:10.3126/nje.v3i2.8512

Kavanaugh, A., Carroll, J. M., Rosson, M. B., Zin, T. T., & Reese, D. (2005). Community Networks: Where offline Communities Meet Online. *Journal of Computer-Mediated Communication*, *10*(4). doi:10.1111/j.1083-6101.2005.tb00266.x

Kavanaugh, A., & Patterson, S. (2001). The Impact of Community Computer Networks on Social Capital and community involvement. *The American Behavioral Scientist*, *45*(3), 496–509. doi:10.1177/00027640121957312

Khanna, S., & Wahi, A. K. (2014). Website Attractiveness in E-Commerce Sites: Key Factors Influencing the Consumer Purchase Decision. *International Journal of Virtual Communities and Social Networking*, *6*(2), 49–59. doi:10.4018/ijvcsn.2014040104

Kietzmann, J. H., Hermkens, K., McCarthy, I. P., & Silvestre, B. S. (2011). Social media? Get serious! Understanding the functional building blocks of social media. *Business Horizons*, *54*(3), 241–251. doi:10.1016/j.bushor.2011.01.005

Kim, J. N., & Rhee, Y. (2011). Strategic thinking about employee communication behavior (ECB) in public relations: Testing the models of megaphoning and scouting effects in Korea. *Journal of Public Relations Research*, *23*(3), 243–268. doi:10.1080/1062726X.2011.582204

Kimmel, A. J., & Kitchen, P. J. (2014). Word of mouth and social media. *Journal of Marketing Communications*, *20*(1-2), 2–4. doi:10.1080/13527266.2013.865868

Kim, S., Zhang, X. A., & Zhang, B. W. (2016). Self-mocking crisis strategy on social media: Focusing on Alibaba chairman Jack Ma in China. *Public Relations Review*, *42*(5), 903–912. doi:10.1016/j.pubrev.2016.10.004

King. (2012). *Candy Crush* [Software]. UK: King.

Koenitz, H. (2014). Five theses for interactive digital narrative. In A. M. C. Fernández-Vara, & D. Thue (Eds.), *Interactive Storytelling: 7th International Conference on Interactive Digital Storytelling (ICIDS 2014)* (Vol. 8832, pp. 134– 139). Berlin, Germany: Springer-Verlag.

Koenitz, H., & Chen, K.-J. (2012). Genres, structures and strategies in interactive digital narratives – analyzing a body of works created in ASAPS. In D. Oyarzun, F. Peinado, R. Young, & A. Gonzalo Méndez (Eds.), *Interactive Storytelling: 5th International Conference on Interactive Digital Storytelling (ICIDS 2012)* (Vol. 7648, pp. 84–95). Berlin, Germany: Springer-Verlag. 10.1007/978-3-642-34851-8_8

Kohut, A. (2012). *Social Networking Popular Across Globe: Arab Publics Most Likely to Express Political Views Online*. Global Attitude Project, PEW Research Center. Retrieved from http://www.pewglobal.org/files/2012/12/Pew-Global-Attitudes-Project-Technology-Report-FINAL-December-12-2012.pdf

Koku, E., Nazer, N., & Wellman, B. (2001). Netting Scholars: Online and Offline. *The American Behavioral Scientist, 44*(10), 1752–1774. doi:10.1177/00027640121958023

Kontaxis, G., Polakis, I., Ioannidis, S., & Markatos, E. P. (2011, March). Detecting social network profile cloning. In *Pervasive Computing and Communications Workshops (PERCOM Workshops), 2011 IEEE International Conference on* (pp. 295-300). IEEE. 10.1109/PERCOMW.2011.5766886

Kumar, R., Novak, J., & Tomkins, A. (2010). Structure and evolution of online social networks. In *Link mining: models, algorithms, and applications* (pp. 337–357). Springer New York. doi:10.1007/978-1-4419-6515-8_13

Kwak, H., Lee, C., Park, H., & Moon, S. (2010, April). What is Twitter, a social network or a news media? In *Proceedings of the 19th international conference on World wide web* (pp. 591-600). ACM. 10.1145/1772690.1772751

Lai, R. K., Tang, J. C., Wong, A. K., & Lei, P. I. (2010). Design and implementation of an online social network with face recognition. *Journal of Advances in Information Technology, 1*(1), 38–42. doi:10.4304/jait.1.1.38-42

Lampe, C. A., Ellison, N., & Steinfield, C. (2007, April). A familiar face (book): profile elements as signals in an online social network. In *Proceedings of the SIGCHI conference on Human factors in computing systems* (pp. 435-444). ACM. 10.1145/1240624.1240695

Lampe, C., Ellison, N., & Steinfield, C. (2006, November). A Face (book) in the crowd: Social searching vs. social browsing. In *Proceedings of the 2006 20th anniversary conference on Computer supported cooperative work* (pp. 167-170). ACM. 10.1145/1180875.1180901

Lange, P. G. (2007). Publicly private and privately public: Social networking on YouTube. *Journal of Computer-Mediated Communication, 13*(1), 361–380. doi:10.1111/j.1083-6101.2007.00400.x

Latane, B. (1981). The Psychology of Social Impact. *The American Psychologist, 36*(4), 343–356. doi:10.1037/0003-066X.36.4.343

Laurel, B. (1986). *Toward the Design of a Computer-based Interactive Fantasy System* (Doctoral dissertation). Ohio State University, Columbus, OH.

Laurel, B. (1991). *Computers as Theatre*. Reading, MA: Addison-Wesley.

Lee, J. (2007). The effects of visual metaphor and cognitive style for mental modeling in a hypermedia- based environment. *Interacting with Computers, 19*(5-6), 614–629. doi:10.1016/j.intcom.2007.05.005

Lee, S. J. (2009). Online Communication and Adolescent Social Ties: Who benefits more from Internet Use? *Journal of Computer-Mediated Communication, 14*(3), 509–531. doi:10.1111/j.1083-6101.2009.01451.x

Lenhart, A., & Madden, M. (2007). *The Use of Social Media Gains a Greater Foothold in Teen Life*. Pew Internet and American Life Project.

Leskovec, J., Huttenlocher, D., & Kleinberg, J. (2010, April). Predicting positive and negative links in online social networks. In *Proceedings of the 19th international conference on World wide web* (pp. 641-650). ACM. 10.1145/1772690.1772756

Liu, H. (2007). Social network profiles as taste performances. *Journal of Computer-Mediated Communication, 13*(1), 252–275. doi:10.1111/j.1083-6101.2007.00395.x

Livingstone, S. (2008). Taking risky opportunities in youthful content creation: Teenagers' use of social networking sites for intimacy, privacy and self-expression. *New Media & Society, 10*(3), 393–411. doi:10.1177/1461444808089415

Louchart, S. (2007). *Emergent Narrative – Towards a Narrative Theory of Virtual Reality* (Doctoral dissertation). University of Salford, Salford, UK.

Louchart, S., & Aylett, R. (2005). Managing a non-linear scenario – a narrative evolution. In G. Subsol (Ed.), *Virtual Storytelling. Using Virtual Reality Technologies for Storytelling. Proceedings of the Third International Conference (ICVS 2005)* (*Vol. 3805*, pp. 148–157). Berlin, Germany: Springer-Verlag. 10.1007/11590361_17

Ludeon Studios. (2016). *RimWorld* [Software]. Canada: Ludeon Studios.

Madejski, M., Johnson, M. L., & Bellovin, S. M. (2011). *The failure of online social network privacy settings*. Academic Press.

Madejski, M., Johnson, M., & Bellovin, S. M. (2011). *The failure of online social network privacy settings*. Department of Computer Science, Columbia University, Tech. Rep. CUCS-010-11.

Mager, R. E. (1988). *The New Six Pack: Making Instruction Work*. Lake.

Magoutis, K., Papoulas, C., Papaioannou, A., Karniavoura, F., Akestoridis, D. G., Parotsidis, N., ... Stephanidis, C. (2015). Design and implementation of a social networking platform for cloud deployment specialists. *Journal of Internet Services and Applications, 6*(1), 19. doi:10.118613174-015-0033-5

Ma, J., Qiao, Y., Hu, G., Huang, Y., Wang, M., Sangaiah, A. K., ... Wang, Y. (2017). Balancing User Profile and Social Network Structure for Anchor Link Inferring Across Multiple Online Social Networks. *IEEE Access: Practical Innovations, Open Solutions, 5*, 12031–12040. doi:10.1109/ACCESS.2017.2717921

Marlow, C., Naaman, M., Boyd, D., & Davis, M. (2006, August). HT06, tagging paper, taxonomy, Flickr, academic article, to read. In *Proceedings of the seventeenth conference on Hypertext and hypermedia* (pp. 31-40). ACM. 10.1145/1149941.1149949

Marshall, G. W., Moncrieff, W. C., & John, M. (2012). Revolution in Sales: The impact of Social media and Related technology on the Selling environment. *Journal of Personal Selling & Sales Management, 32*(3), 349–363. doi:10.2753/PSS0885-3134320305

Mateas, M. (2002). *Interactive Drama, Art and Artificial Intelligence* (Doctoral dissertation). Carnegie Mellon University, Pittsburgh, PA.

Mateas, M., & Stern, A. (2002). *Architecture, authorial idioms and early observations of the interactive drama Façade*. Technical Report CMU-CS-02-198, School of Computer Science, Carnegie Mellon University, Pittsburgh, PA.

Mateas, M., & Stern, A. (2005). *Façade* [Software]. USA: Procedural Arts. Available from http://www.interactivestory.net/

Mautone, P. D., & Mayer, R. E. (2001). Signaling as a cognitive guide in multimedia learning. *Journal of Educational Psychology, 93*(2), 377–389. doi:10.1037/0022-0663.93.2.377

Maxis. (2000). *The Sims* [Software]. USA: Electronic Arts.

Mayer, R. E. (2001). *Multimedia Learning*. Cambridge, UK: Cambridge University Press. doi:10.1017/CBO9781139164603

McKay, E. (2008). The Human-Dimensions of Human-Computer Interaction: Balancing the HCI Equation (Vol. 3). Amsterdam: IOS Press.

McKay, E., & Izard, J. (2013). Seamless Web-Mediated Training Courseware Design Model: Innovating Adaptive Educational-Learning Systems. In A. P. Ayala (Ed.), *Intelligent and Adaptive Educational-Learning Systems: Achievements and Trends* (Vol. 17, pp. 417-442). Springer Berlin Heidelberg.

McKay, E., & Izard, J. (2016). *Planning effective HCI courseware design to enhance online education and training*. Paper presented at the 18th International Conference on Universal Access in Human-Computer Interaction, Toronto, Canada. 10.1007/978-3-319-39399-5_18

McKay, E., & Izard, J. F. (2014). *Online Training Design: Workforce reskilling in Government agencies*. Paper presented at the IC3e 2014 IEEE Conference on eLearning, eManagement and Services, Swinburne University, Melbourne, Australia. 10.1109/IC3e.2014.7081244

McKay, E., & Izard, J. F. (2015). *eLearning programme design? Customised for user-centrered participation, Virtual paper presented (and published in Proceedings)*. Paper presented at the Global Learn 2015: Global Conference on Learning and Technology, Berlin, Germany.

McKay, E., & Izard, J. F. (2015a). *Evaluate Online Training Effectiveness: Differentiate What They Do and Do Not Know*. Paper presented at the 8th International Conference on ICT, Society and Human Beings 2015 (Multi conference on computer science and information systems - MCCSIS), Las Palmas de Gran Canaria, Spain.

McKay, E., & Izard, J. F. (2015b). *Measurement of cognitive performance in introductory ethics: Workforce reskilling in government agencies*. Paper presented at the 4th IEEE Conference on e-Learning, e-Management and e-Services (IC3e 2015), Melaka, Malaysia. 10.1109/IC3e.2015.7403478

McKay, E., & Vilela, C. (2011). Online Workforce Training in Government Agencies: Towards effective instructional systems design. Australian Journal of Adult Learning, 51(2), 302-328.

McKay, E., Axmann, M., Banjanin, N., & Howat, A. (2007). *Towards Web-Mediated Learning Reinforcement: Rewards for Online Mentoring Through Effective Human-Computer Interaction.* Paper presented at the 6th IASTED International Conference on Web-Based Education, Chamonix, France. Retrieved 15/04/07 http://www.iasted.org/conferences/pastinfo-557.html

McKay, E. (2018). Prescriptive Training Courseware: IS-Design Methodology. *AJIS. Australasian Journal of Information Systems, 22.*

McKeen, J. (2011). *Making IT Happen: Critical Issues in IT Management.* Wiley Series in Information Systems.

McNurlin, B. (2009). *Information Systems Management in Practice.* Prentice Hall.

McQuail, D. (2000). *McQuail's Mass Communication Theory.* London: Sage.

McQuail, D., Blumler, J. G., & Brown, J. (1972). The television audience: A revised perspective. In D. McQuail (Ed.), *Sociology of Mass Communication* (pp. 65–135). Middlesex, UK: Penguin.

Mellenbergh, G. (2008). *Advising on research methods: A Consultant's Companion.* Johannes van Kessel Publishing.

Men, L. R. (2014). Why leadership matters to internal communication: Linking transformational leadership, symmetrical communication, and employee outcomes. *Journal of Public Relations Research, 26*(3), 256–279. doi:10.1080/1062726X.2014.908719

Men, L. R. (2015). The internal communication role of the chief executive officer: Communication channels, style, and effectiveness. *Public Relations Review, 41*(4), 461–471. doi:10.1016/j.pubrev.2015.06.021

Men, L. R., & Stacks, D. (2014). The effects of authentic leadership on strategic internal communication and employee-organization relationships. *Journal of Public Relations Research, 26*(4), 301–324. doi:10.1080/1062726X.2014.908720

Men, L. R., & Tsai, W. H. S. (2016). Public engagement with CEOs on social media: Motivations and relational outcomes. *Public Relations Review, 42*(5), 932–942. doi:10.1016/j.pubrev.2016.08.001

Merleau-Ponty, M. (1964). *Signs* (R. McCleary, Trans.). Evanston, IL: Northwestern University Press.

Merrill, M. D. (2002b). First Principles of Instruction. *ETR&D, 50*(3), 43-59. Available from http://www.indiana.edu/~tedfrick/aect2002/firstprinciplesbymerrill.pdf

Merrill, M. D. (2002c). Pebble-in-the-Pond Model for Instructional Development. *Performance Measurement, 41*(7), 39-44. Available from http://mdavidmerrill.com/Papers/Pebble_in_the_Pond.pdf

Merrill, D. M., Barclay, M., & Van-Schaak, A. (2008). Prescriptive Principles for Instructional Design. In J. M. Spector, M. D. Merrill, M. F. J. van Merriënboer, & M. P. Driscoll (Eds.), *Foundations of Educational Technology: Integrative approaches and interdisciplinary perspectives* (pp. 173–185). Taylor & Frances.

Merrill, D. M., Barclay, M., & Van-Schaak, A. (2016). Prescriptive Principles for Instructional Design. In J. M. Spector (Ed.), *Foundations of Educational Technology: Integrative Approaches and Interdisciplinary Perspectives*. Taylor & Frances.

Merrill, M. (2002). First principles of instruction. *Educational Technology Research and Development*, *50*(3), 43–59. doi:10.1007/BF02505024

Merrill, M. D. (2002a). Effective Use of Instructional Technology Requires Educational Reform. *Educational Technology*, *42*(4), 13–16.

Metzger, M. J., Flanagin, A. J., & Medders, R. B. (2010). Social and heuristic approaches to credibility evaluation online. *Journal of Communication*, *60*(3), 413–439. doi:10.1111/j.1460-2466.2010.01488.x

Minor, J.I. (2014). *Using Knowledge of Transactional Distance Theory to Strengthen an Online Developmental Reading Course: An action research study* (PhD dissertation). Capella. (3614806)

Miranda, F.J., Rubio, S., & Chamorro-Mera A. (2014). Facebook as a Marketing Tool: An Analysis of the 100 Top-Ranked Global Brands. *International Journal of Virtual Communities and Social Networking Archive*, *6*(4), 14-28.

Mir, I., & Zaheer, A. (2012). Verification of social impact theory claims in social media context. *Journal of Internet Banking and Commerce*, *17*(1), 1–15.

Mislove, A., Koppula, H. S., Gummadi, K. P., Druschel, P., & Bhattacharjee, B. (2008, August). Growth of the flickr social network. In *Proceedings of the first workshop on Online social networks* (pp. 25-30). ACM. 10.1145/1397735.1397742

Mislove, A., Marcon, M., Gummadi, K. P., Druschel, P., & Bhattacharjee, B. (2007, October). Measurement and analysis of online social networks. In *Proceedings of the 7th ACM SIGCOMM conference on Internet measurement* (pp. 29-42). ACM. 10.1145/1298306.1298311

Mislove, A., Viswanath, B., Gummadi, K. P., & Druschel, P. (2010, February). You are who you know: inferring user profiles in online social networks. In *Proceedings of the third ACM international conference on Web search and data mining* (pp. 251-260). ACM. 10.1145/1718487.1718519

Mohamad, M, & McKay, E. (2014). Measuring the Effects of Cognitive Preference to Enhance Online Instruction Through Sound ePedagogy Design. *International Journal of Technology Enhanced Learning*, *7*(1), 21-25.

Mohamad, M. (2012). *The Effects of Web-Mediated Instructional Strategies and Cognitive Preferences in the Acquisition of Introductory Programming Concepts: A Rasch Model Approach*. RMIT University, School of Business Information Technology and Logistics.

Mourtada, R., & Salem, F. (2011). Civil movements: The Impact of Facebook and Twitter. *Arab Social Media Report, 1*(2), 1–30.

Mukherjee, S. (2016). Leadership network and team performance in interactive contests. *Social Networks, 47*, 85–92. doi:10.1016/j.socnet.2016.05.003

Muntinga, D. G., Moorman, M., & Smit, E. G. (2011). Introducing COBRAs: Exploring motivations for brand-related social media use. *International Journal of Advertising, 30*(1), 13–46. doi:10.2501/IJA-30-1-013-046

Murray, J. H. (1997). *Hamlet on the Holodeck: The Future of Narrative in Cyberspace.* Cambridge, MA: The MIT Press.

Murray, J. H. (2012). *Inventing the Medium: Principles of Interactions Design as a Cultural Practice.* Cambridge, MA: The MIT Press.

Narendula, R., Papaioannou, T. G., & Aberer, K. (2011, August). My3: A highly-available P2P-based online social network. In *Peer-to-Peer Computing (P2P), 2011 IEEE International Conference on*(pp. 166-167). IEEE.

Nie, N. H. (2001). Sociability, Interpersonal Relations, and the Internet: Reconciling Conflicting Findings. *The American Behavioral Scientist, 45*(3), 420–435. doi:10.1177/00027640121957277

Nie, Y., Jia, Y., Li, S., Zhu, X., Li, A., & Zhou, B. (2016). Identifying users across social networks based on dynamic core interests. *Neurocomputing, 210*, 107–115. doi:10.1016/j. neucom.2015.10.147

Nosko, A., Wood, E., & Molema, S. (2010). All about me: Disclosure in online social networking profiles: The case of FACEBOOK. *Computers in Human Behavior, 26*(3), 406–418. doi:10.1016/j. chb.2009.11.012

Nowak, A., Szamrejand, J., & Latane, B. (1990). From Private Attitude to Public Opinion: A Dynamic Theory of Social Impact. *Psychological Review, 97*(3), 362–376. doi:10.1037/0033-295X.97.3.362

Nunnaly, J. (1978). *Psychometric theory.* New York: McGraw-Hill.

O'Brien, J. (1999). *Management Information Systems: Managing Information Technology in the Internet-worked Enterprise.* Boston: Irwin McGraw-Hill.

O'Reilly, K., & Marx, S. (2011). How young, technical consumers assess online WOM credibility. *Qualitative Market Research, 14*(4), 330–359. doi:10.1108/13522751111163191

Panofsky, E. (2003). *Iconography and Iconology: An Introduction to the Study of Renaissance Art.* Chicago, IL: University of Chicago Press.

Papacharissi, Z. (2009). The virtual geographies of social networks: A comparative analysis of Facebook, LinkedIn and ASmallWorld. *New Media & Society, 11*(1-2), 199–220. doi:10.1177/1461444808099577

Parker, N. K. (2004). The Quality Dilemma in Online Education. In T. Anderson & F. Elloumi (Eds.), *Theory and Practice of Online Learning* (pp. 385 - 409). Athabasca University. Retrieved 11/05/06 http://cde.athabascau.ca/online_book/copyright.html

Park, J., Kim, H., Cha, M., & Jeong, J. (2011). *Ceo's apology in twitter: A case study of the fake beef labeling incident by e-mart*. Springer Berlin Heidelberg.

Park, N., Kee, K. F., & Valenzuela, S. (2009). Being immersed in social networking environment: Facebook groups, uses and gratifications, and social outcomes. *Cyberpsychology & Behavior*, *12*(6), 729–733. doi:10.1089/cpb.2009.0003 PMID:19619037

Pempek, T. A., Yermolayeva, Y. A., & Calvert, S. L. (2009). College students' social networking experiences on Facebook. *Journal of Applied Developmental Psychology*, *30*(3), 227–238. doi:10.1016/j.appdev.2008.12.010

Pereira, R., Baranauskas, M. C. C., & da Silva, S. R. P. (2010, June). Social software building blocks: revisiting the honeycomb framework. In *Information Society (i-Society), 2010 International Conference on* (pp. 253-258). IEEE.

Phua, J. J., & Jin, S. A. (2011). 'Finding a Home Away From Home': The Use of Social Networking Sites by Asia-Pacific Students in the United States for Bridging and Bonding Social Capital. *Asian Journal of Communication*, *21*(5), 504–519. doi:10.1080/01292986.2011.587015

Pjero, E., & Kercini, D. (2015). Social media and consumer behavior – How does it works in albania reality? *Academic Journal of Interdisciplinary Studies*, *4*(3), 141–146.

Plotnick, L., White, C., & Plummer, M. M. (2009). The design of an online social network site for emergency management: A one stop shop. *AMCIS 2009 Proceedings*, 420.

Porter, M. C., Anderson, B., & Nhotsavang, M. (2015). Anti-social media: Executive Twitter "engagement" and attitudes about media credibility. *Journal of Communication Management*, *19*(3), 270–287. doi:10.1108/JCOM-07-2014-0041

Putnam, R. D. (2000). *Bowling Alone: The Collapse and Revival of American Community*. New York: Simon & Schuster. doi:10.1145/358916.361990

Raacke, J., & Bonds-Raacke, J. (2008). MySpace and Facebook: Applying the uses and gratifications theory to exploring friend-networking sites. *Cyberpsychology & Behavior*, *11*(2), 169–174. doi:10.1089/cpb.2007.0056 PMID:18422409

Rathore, A. K., & Ilavarasan, P. V. (2018). Social Media and Business Practices. In Encyclopedia of Information Science and Technology, Fourth Edition (pp. 7126-7139). IGI Global. doi:10.4018/978-1-5225-2255-3.ch619

Rathore, A. K., Ilavarasan, P. V., & Dwivedi, Y. K. (2016). Social media content and product co-creation: An emerging paradigm. *Journal of Enterprise Information Management*, *29*(1), 7–18. doi:10.1108/JEIM-06-2015-0047

Red Barrels. (2017). *Outlast 2* [Software]. Canada: Red Barrels.

Regnell, B., & Brinkkemper, S. (2005). *13 Market-Driven Requirements Engineering for Software Products*. Academic Press.

Report - 2013 Social CEO Report: Are America's Top CEOs Getting More Social? (n.d.). Domo. Retrieved March 3, 2014, from https://www.domo.com/learn/2013-social-ceo-report-are-americas-top-ceos-getting-more-social

Resick, C. J., Whitman, D. S., Weingarden, S. M., & Hiller, N. J. (2009). The bright-side and the dark-side of CEO personality: Examining core self-evaluations, narcissism, transformational leadership, and strategic influence. *The Journal of Applied Psychology*, *94*(6), 1365–1381. doi:10.1037/a0016238 PMID:19916649

Rezaee, S., Lavesson, N., & Johnson, H. (2012). E-mail prioritization using online social network profile distance. *Computer Science & Applications*, *9*(1), 70–87.

Riding, R. J. (2005). *CSA Making Learning Effective - Cognitive Style and Effective Learning*. Learning & Training Technology.

Riding, R., & Douglas, G. (1993). The effect of cognitive style and mode of presentation on learning performance. *The British Journal of Educational Psychology*, *63*(2), 297–307. doi:10.1111/j.2044-8279.1993.tb01059.x PMID:8353062

Ridings, C.M. & Gefen, D. (2004). Virtual community attraction: why people hang out online. *Journal of Computer-Mediated Communication, 10*(1), 1-10.

Riedl, M. (2004). *Narrative Generation: Balancing Plot and Character* (Doctoral dissertation). North Carolina State University, Raleigh, NC.

Robson, C. (2007). *Real-world research: A resource for social scientists and practitioner researchers*. Malden, MA: Blackwell Publishing.

Ryan, M.-L. (2001). *Narrative as Virtual Reality: Immersion and Interactivity in Literature and Electronic Media*. Baltimore, MD: Johns Hopkins University Press.

Rybalko, S., & Seltzer, T. (2010). Dialogic communication in 140 characters or less: How Fortune 500 companies engage stakeholders using Twitter. *Public Relations Review*, *36*(4), 336–341. doi:10.1016/j.pubrev.2010.08.004

Salem, F. (2009). *Cross-Agency Collaboration in the UAE Government: The Role of Trust and the Impact of Technology*. Retrieved October 22, 2014 from http://www.mbrsg.ae/HOME/PUBLICATIONS/Research-Report-Research-Paper-White-Paper/Cross-Agency-Collaboration-in-the-UAE-Government-T.aspx

Saluja, D., & Singh, S. (2014). Impact of Social Media Marketing Strategies on Consumers Behaviour in Delhi. *International Journal of Virtual Communities and Social Networking*, *6*(2), 1–23. doi:10.4018/ijvcsn.2014040101

Schapiro, M. (1969). On Some Problems in the Semiotics of Visual Art: Field and Vehicle in Image-Signs. *Semiotica*, *I*, 223–242.

Schmidt, G. (Producer), & Hirschbiegel, O. (Director). (1991, December 15). *Mörderische Entscheidung* [Television broadcast]. Germany: Das Erste & ZDF.

Schneider, F., Feldmann, A., Krishnamurthy, B., & Willinger, W. (2009, November). Understanding online social network usage from a network perspective. In *Proceedings of the 9th ACM SIGCOMM conference on Internet measurement conference* (pp. 35-48). ACM. 10.1145/1644893.1644899

Schultz, F., Utz, S., & Göritz, A. (2011). Is the medium the message? Perceptions of and reactions to crisis communication via twitter, blogs and traditional media. *Public Relations Review*, *37*(1), 20–27. doi:10.1016/j.pubrev.2010.12.001

Sheth, S., Kaiser, G., & Maalej, W. (2014, May). Us and them: a study of privacy requirements across North America, Asia, and Europe. In *Proceedings of the 36th International Conference on Software Engineering* (pp. 859-870). ACM. 10.1145/2568225.2568244

Silfverberg, S., Liikkanen, L. A., & Lampinen, A. (2011, March). I'll press play, but I won't listen: profile work in a music-focused social network service. In *Proceedings of the ACM 2011 conference on Computer supported cooperative work* (pp. 207-216). ACM. 10.1145/1958824.1958855

Singh, R. R., & Tomar, D. S. (2009). *Approaches for user profile investigation in orkut social network*. arXiv preprint arXiv:0912.1008

Singh, S. (2016). Role of Social Media Marketing Strategies on Customer Perception. *Anveshak-International Journal Of Management*, *5*(2), 27–41. doi:10.15410/aijm/2016/v5i2/100695

Skeels, M. M., & Grudin, J. (2009, May). When social networks cross boundaries: a case study of workplace use of facebook and linkedin. In *Proceedings of the ACM 2009 international conference on Supporting group work* (pp. 95-104). ACM. 10.1145/1531674.1531689

Smed, J. (2014). Interactive storytelling: Approaches, applications, and aspirations. *International Journal of Virtual Communities and Social Networking*, *6*(1), 22–34. doi:10.4018/ijvcsn.2014010102

Smed, J., & Hakonen, H. (2017). *Algorithms and Networking for Computer Games* (2nd ed.). Chichester, UK: John Wiley & Sons. doi:10.1002/9781119259770

Spierling, U. (2009). Conceiving interactive story events. In I.A. Iurgel, N. Zagalo, & P. Petta (Eds.), *Interactive Storytelling: Second Joint Conference on Interactive Digital Storytelling (ICIDS 2009)* (*Vol. 5915*, pp. 292–297). Berlin, Germany: Springer-Verlag.

Spierling, U., & Szilas, N. (2009). Authoring issues beyond tools. In I.A. Iurgel, N. Zagalo, & P. Petta (Eds.), *Interactive Storytelling: Second Joint Conference on Interactive Digital Storytelling (ICIDS 2009)* (*Vol. 5915*, pp. 50–61). Berlin, Germany: Springer-Verlag.

Stark, H. T., & Krosnick, A. J. (2017). GENSI: A new graphical tool to collect ego-centered network data react. *Social Networks*, *48*, 36–45. doi:10.1016/j.socnet.2016.07.007

Steyn, B. (2004). From strategy to corporate communication strategy: A conceptualisation. *Journal of Communication Management*, *8*(2), 168–183. doi:10.1108/13632540410807637

Strater, K., & Lipford, H. R. (2008, September). Strategies and struggles with privacy in an online social networking community. In *Proceedings of the 22nd British HCI Group Annual Conference on People and Computers: Culture, Creativity, Interaction* (vol. 1, pp. 111-119). British Computer Society.

Strater, K., & Lipford, H. R. (2008, September). Strategies and struggles with privacy in an online social networking community. *Interaction, 1*, 111–119.

Strufe, T. (2010, April). Profile popularity in a business-oriented online social network. In *Proceedings of the 3rd workshop on social network systems* (p. 2). ACM. 10.1145/1852658.1852660

Subrahmanyam, K., Reich, S. M., Waechter, N., & Espinoza, G. (2008). Online and offline social networks: Use of social networking sites by emerging adults. *Journal of Applied Developmental Psychology, 29*(6), 420–433. doi:10.1016/j.appdev.2008.07.003

Sun, S.-Y., Ju, T. L., Chumg, H.-F., Wu, C.-Y., & Chao, P.-J. (2009). Influence on Willingness of Virtual Community's Knowledge Sharing: Based on Social Capital Theory and Habitual Domain. *World Academy of Science, Engineering and Technology, 53*, 142–149.

Supergiant Games. (2011). *Bastion* [Software]. USA: Warner Bros. Interactive Entertainment.

Swedowsky, M. (2009). *Improving Customer Experience by Listening and Responding to Social Media.* Retrieved September 8, 2013, from http://blog.nielsen.com/ nielsenwire/consumer/ improving-customer-experienceby-listening- and-responding-to-social-media/

Szilas, N., Barles, J., & Kavakli, M. (2007). An implementation of real-time 3D interactive drama. *Computers in Entertainment, 5*(1), 5. doi:10.1145/1236224.1236233

The Arab World Online Report: Trends in Internet Usage in the Arab Region. (2013). Dubai School of Government.

The State of Broadband 2014: Broadband for all. (2014). A report by the Broadband Commission, United Nations: Educational, Scientific, Cultural Organization.

Tice, D. M., Butler, J. L., Muraven, M. B., & Stillwell, A. M. (1995). When Modesty Prevails: Differential Favorability of Self-Presentation to Friends and Strangers. *Journal of Personality and Social Psychology, 69*(6), 1120–1138. doi:10.1037/0022-3514.69.6.1120

Toor, A., Husnain, M., & Hussain, T. (2017). The Impact of Social Network Marketing on Consumer Purchase Intention in Pakistan: Consumer Engagement as a Mediator. *Asian Journal of Business and Accounting, 10*(1), 167–199.

Trusov, M., Bucklin, R., & Pauwels, K. (2009). Effects of Word-of-Mouth Versus Traditional Marketing: Findings from an Internet Social Networking Site. *Journal of Marketing, 73*(September), 90–102. doi:10.1509/jmkg.73.5.90

United Arab Emirates' National Bureau of Statistics. (2010). Retrieved on 24th January 2015, from: http://www.uaestatistics.gov.ae/ReportPDF/Population%20Estimates%202006%20-%202010.pdf

Utz, S. (2010). Show me your friends and I will tell you what type of person you are: How one's profile, number of friends, and type of friends influence impression formation on social network sites. *Journal of Computer-Mediated Communication, 15*(2), 314–335. doi:10.1111/j.1083-6101.2010.01522.x

Utz, S., & Beukeboom, C. J. (2011). The role of social network sites in romantic relationships: Effects on jealousy and relationship happiness. *Journal of Computer-Mediated Communication, 16*(4), 511–527. doi:10.1111/j.1083-6101.2011.01552.x

Valenzuela, S., Park, N., & Kee, K. F. (2009). Is There Social Capital in a Social Network Site?: Facebook Use and College Students' Life Satisfaction, Trust, and Participation1. *Journal of Computer-Mediated Communication, 14*(4), 875–901. doi:10.1111/j.1083-6101.2009.01474.x

Valve. (1998). *Half-Life* [Software]. USA: Sierra Studios.

Van Der Sype, Y. S., & Maalej, W. (2014, August). On lawful disclosure of personal user data: What should app developers do? In *Requirements Engineering and Law (RELAW), 2014 IEEE 7th International Workshop on* (pp. 25-34). IEEE.

Van Lamsweerde, A. (2007). *Requirements engineering*. John Wiley & Sons.

Van Lamsweerde, A. (2008, November). Requirements engineering: from craft to discipline. In *Proceedings of the 16th ACM SIGSOFT International Symposium on Foundations of software engineering* (pp. 238-249). ACM. 10.1145/1453101.1453133

Van Tubergen, F., Al-Modaf, O. A., Almosaed, N. F., & Al-Ghamdi, M. S. (2016). Personal networks in Saudi Arabia: The role of ascribed and achieved characteristics. *Social Networks, 45*, 45–54. doi:10.1016/j.socnet.2015.10.007

Versu. (2013). *Versu* [Software]. USA: Versu. Available from https://versu.com/

Vidgen, R., Mark Sims, J., & Powell, P. (2013). Do CEO bloggers build community? *Journal of Communication Management, 17*(4), 364–385. doi:10.1108/JCOM-08-2012-0068

Villegas, J. (2011). The Influence of Social Media Usage and Online Social Capital on Advertising Perception. *American Academy of Advertising Conference Proceedings*, 69.

Vinerean, S., Cetina, I., & Tichindelean, M. (2013). The effects of social media marketing on online consumer behavior. *International Journal of Business and Management, 8*(14), 66–69. doi:10.5539/ijbm.v8n14p66

Viswanath, B., Mislove, A., Cha, M., & Gummadi, K. P. (2009, August). On the evolution of user interaction in facebook. In *Proceedings of the 2nd ACM workshop on Online social networks* (pp. 37-42). ACM. 10.1145/1592665.1592675

Vitak, J. (2012). The Impact of Context Collapse and Privacy on Social Network Site Disclosures. *Journal of Broadcasting & Electronic Media, 56*(4), 451–470. doi:10.1080/08838151.2012.732140

Von Hippel, E. (2005). *Innovation*. Cambridge, MA: MIT Press.

Walkins, S. C., & Lee, H. E. (2009). *Bonding, Bridging and Friending: Investigating the Social Aspects of Social Networking Sites.* Paper presented in the 95th Annual Convention of the National Communication Association (NCA), Chicago. IL.

Wardrip-Fruin, N. (2009). *Expressive Processing: Digital Fictions, Computer Games and Software Studies.* Cambridge, MA: The MIT Press.

Waters, R. D., Burnett, E., Lamm, A., & Lucas, J. (2009). Engaging stakeholders through social networking: How nonprofit organizations are using Facebook. *Public Relations Review, 35*(2), 102–106. doi:10.1016/j.pubrev.2009.01.006

Watkins, B. A. (2017). Experimenting with dialogue on Twitter: An examination of the influence of the dialogic principles on engagement, interaction, and attitude. *Public Relations Review, 43*(1), 163–171. doi:10.1016/j.pubrev.2016.07.002

Weallans, A., Louchart, S., & Aylett, R. (2012). Distributed drama management: Beyond double appraisal in emergent narrative. In D. Oyarzun, F. Peinado, R. Young, & A. Gonzalo Méndez (Eds.), *Interactive Storytelling: 5th International Conference on Interactive Digital Storytelling (ICIDS 2012)* (*Vol. 7648*, pp. 132–143). Berlin, Germany: Springer-Verlag. 10.1007/978-3-642-34851-8_13

Weber Shandwick. (2012). *The social CEO: executives tell all.* Retrieved from. http://www.webershandwick.com/uploads/news/files/Social-CEO-Study.pdf

Weiss, J. B., Berner, E. S., Johnson, K. B., Giuse, D. A., Murphy, B. A., & Lorenzi, N. M. (2013). Recommendations for the design, implementation and evaluation of social support in online communities, networks, and groups. *Journal of Biomedical Informatics, 46*(6), 970–976. doi:10.1016/j.jbi.2013.04.004 PMID:23583424

Wellman, B., Haase, A. Q., Witte, J., & Hampton, K. (2010). Does the Internet Increase, Decrease or Supplement Social Capital? Social Networks, Participation and Community Commitment. *The American Behavioral Scientist, 45*(3), 436–455. doi:10.1177/00027640121957286

Wiest, J., & Eltantawy, N. (2012). Social media use among UAE College Students One Year after the Arab Spring. *Journal of Arab and Muslim Media Research, 5*(3), 209–226. doi:10.1386/jammr.5.3.209_1

Wikle, T. A., & Comer, J. C. (2012). Facebook's Rise to the Top: Exploring Social Networking Registrations by Country. *International Journal of Virtual Communities and Social Networking, 4*(2), 46–60. doi:10.4018/jvcsn.2012040104

Williams, D. (2006). On and Off the 'Net: Scales for Social Capital in an Online Era. *Journal of Computer-Mediated Communication, 11*(2), 593–628. doi:10.1111/j.1083-6101.2006.00029.x

Williams, K. D., & Williams, K. B. (1989). Impact of source strength on two compliance techniques. *Basic and Applied Social Psychology, 10*(2), 149–159. doi:10.120715324834basp1002_5

Wischenbart, M., Mitsch, S., Kapsammer, E., Kusel, A., Pröll, B., Retschitzegger, W., ... Lechner, S. (2012, April). User profile integration made easy: model-driven extraction and transformation of social network schemas. In *Proceedings of the 21st International Conference on World Wide Web* (pp. 939-948). ACM. 10.1145/2187980.2188227

Wong, W. K. W. (2012). Faces on Facebook: A Study of Self-Presentation and Social Support on Facebook. *Discovery-SS Student E-Journal, 1*, 184–214.

Wu, M., & Adams, R. (2007). *Applying the Rasch Model to Psycho-social Measurement: A practical Approach*. Academic Press.

Xiang, R., Neville, J., & Rogati, M. (2010, April). Modeling relationship strength in online social networks. In *Proceedings of the 19th international conference on World wide web* (pp. 981-990). ACM. 10.1145/1772690.1772790

Yang, T. (2012). The decision behaviour of Facebook users. *Journal of Computer Information Systems, 52*(3), 50–59.

Young, P. A. (2008). Integrating culture in the design of ICTs. *British Journal of Educational Technology, 39*(1), 6–17.

Yu, E. S. (1997, January). Towards modelling and reasoning support for early-phase requirements engineering. In *Requirements Engineering, 1997., Proceedings of the Third IEEE International Symposium on* (pp. 226-235). IEEE. 10.1109/ISRE.1997.566873

Yu-Chen, H. (2007). The Effects of Visual Versus Verbal Metaphors on Novice and Expert Learners' Performance. In J. A. Jacko (Ed.), *Human-Computer Interaction Part IV HCII* (pp. 264–269). Berlin: Springer-Verlag.

Zhang, H., Kan, M., Liu, Y., & Ma, S. (2014, November). Online social network profile linkage based on cost-sensitive feature acquisition. In *Chinese National Conference on Social Media Processing* (pp. 117-128). Springer Berlin Heidelberg. 10.1007/978-3-662-45558-6_11

Zhang, H., Kan, M. Y., Liu, Y., & Ma, S. (2014, December). Online social network profile linkage. In *Asia Information Retrieval Symposium* (pp. 197-208). Springer.

Zhang, W., Nie, L., Jiang, H., Chen, Z., & Liu, J. (2014). Developer social networks in software engineering: Construction, analysis, and applications. *Science China. Information Sciences, 57*(12), 1–23.

Zhou, X., Liang, X., Zhang, H., & Ma, Y. (2016). Cross-platform identification of anonymous identical users in multiple social media networks. *IEEE Transactions on Knowledge and Data Engineering, 28*(2), 411–424. doi:10.1109/TKDE.2015.2485222

Related References

To continue our tradition of advancing information science and technology research, we have compiled a list of recommended IGI Global readings. These references will provide additional information and guidance to further enrich your knowledge and assist you with your own research and future publications.

Adesina, K., Ganiu, O., & R., O. S. (2018). Television as Vehicle for Community Development: A Study of Lotunlotun Programme on (B.C.O.S.) Television, Nigeria. In A. Salawu, & T. Owolabi (Eds.), *Exploring Journalism Practice and Perception in Developing Countries* (pp. 60-84). Hershey, PA: IGI Global. doi:10.4018/978-1-5225-3376-4.ch004

Adigun, G. O., Odunola, O. A., & Sobalaje, A. J. (2016). Role of Social Networking for Information Seeking in a Digital Library Environment. In A. Tella (Ed.), *Information Seeking Behavior and Challenges in Digital Libraries* (pp. 272–290). Hershey, PA: IGI Global. doi:10.4018/978-1-5225-0296-8.ch013

Ahmad, M. B., Pride, C., & Corsy, A. K. (2016). Free Speech, Press Freedom, and Democracy in Ghana: A Conceptual and Historical Overview. In L. Mukhongo & J. Macharia (Eds.), *Political Influence of the Media in Developing Countries* (pp. 59–73). Hershey, PA: IGI Global. doi:10.4018/978-1-4666-9613-6.ch005

Ahmad, R. H., & Pathan, A. K. (2017). A Study on M2M (Machine to Machine) System and Communication: Its Security, Threats, and Intrusion Detection System. In M. Ferrag & A. Ahmim (Eds.), *Security Solutions and Applied Cryptography in Smart Grid Communications* (pp. 179–214). Hershey, PA: IGI Global. doi:10.4018/978-1-5225-1829-7.ch010

Akanni, T. M. (2018). In Search of Women-Supportive Media for Sustainable Development in Nigeria. In A. Salawu & T. Owolabi (Eds.), *Exploring Journalism Practice and Perception in Developing Countries* (pp. 126–149). Hershey, PA: IGI Global. doi:10.4018/978-1-5225-3376-4.ch007

Akçay, D. (2017). The Role of Social Media in Shaping Marketing Strategies in the Airline Industry. In V. Benson, R. Tuninga, & G. Saridakis (Eds.), *Analyzing the Strategic Role of Social Networking in Firm Growth and Productivity* (pp. 214–233). Hershey, PA: IGI Global. doi:10.4018/978-1-5225-0559-4.ch012

Al-Rabayah, W. A. (2017). Social Media as Social Customer Relationship Management Tool: Case of Jordan Medical Directory. In W. Al-Rabayah, R. Khasawneh, R. Abu-shamaa, & I. Alsmadi (Eds.), *Strategic Uses of Social Media for Improved Customer Retention* (pp. 108–123). Hershey, PA: IGI Global. doi:10.4018/978-1-5225-1686-6.ch006

Almjeld, J. (2017). Getting "Girly" Online: The Case for Gendering Online Spaces. In E. Monske & K. Blair (Eds.), *Handbook of Research on Writing and Composing in the Age of MOOCs* (pp. 87–105). Hershey, PA: IGI Global. doi:10.4018/978-1-5225-1718-4.ch006

Altaş, A. (2017). Space as a Character in Narrative Advertising: A Qualitative Research on Country Promotion Works. In R. Yılmaz (Ed.), *Narrative Advertising Models and Conceptualization in the Digital Age* (pp. 303–319). Hershey, PA: IGI Global. doi:10.4018/978-1-5225-2373-4.ch017

Altıparmak, B. (2017). The Structural Transformation of Space in Turkish Television Commercials as a Narrative Component. In R. Yılmaz (Ed.), *Narrative Advertising Models and Conceptualization in the Digital Age* (pp. 153–166). Hershey, PA: IGI Global. doi:10.4018/978-1-5225-2373-4.ch009

An, Y., & Harvey, K. E. (2016). Public Relations and Mobile: Becoming Dialogic. In X. Xu (Ed.), *Handbook of Research on Human Social Interaction in the Age of Mobile Devices* (pp. 284–311). Hershey, PA: IGI Global. doi:10.4018/978-1-5225-0469-6.ch013

Assay, B. E. (2018). Regulatory Compliance, Ethical Behaviour, and Sustainable Growth in Nigeria's Telecommunications Industry. In I. Oncioiu (Ed.), *Ethics and Decision-Making for Sustainable Business Practices* (pp. 90–108). Hershey, PA: IGI Global. doi:10.4018/978-1-5225-3773-1.ch006

Averweg, U. R., & Leaning, M. (2018). The Qualities and Potential of Social Media. In M. Khosrow-Pour, D.B.A. (Ed.), Encyclopedia of Information Science and Technology, Fourth Edition (pp. 7106-7115). Hershey, PA: IGI Global. doi:10.4018/978-1-5225-2255-3.ch617

Azemi, Y., & Ozuem, W. (2016). Online Service Failure and Recovery Strategy: The Mediating Role of Social Media. In W. Ozuem & G. Bowen (Eds.), *Competitive Social Media Marketing Strategies* (pp. 112–135). Hershey, PA: IGI Global. doi:10.4018/978-1-4666-9776-8.ch006

Baarda, R. (2017). Digital Democracy in Authoritarian Russia: Opportunity for Participation, or Site of Kremlin Control? In R. Luppicini & R. Baarda (Eds.), *Digital Media Integration for Participatory Democracy* (pp. 87–100). Hershey, PA: IGI Global. doi:10.4018/978-1-5225-2463-2.ch005

Bacallao-Pino, L. M. (2016). Radical Political Communication and Social Media: The Case of the Mexican #YoSoy132. In T. Deželan & I. Vobič (Eds.), *R)evolutionizing Political Communication through Social Media* (pp. 56–74). Hershey, PA: IGI Global. doi:10.4018/978-1-4666-9879-6.ch004

Baggio, B. G. (2016). Why We Would Rather Text than Talk: Personality, Identity, and Anonymity in Modern Virtual Environments. In B. Baggio (Ed.), *Analyzing Digital Discourse and Human Behavior in Modern Virtual Environments* (pp. 110–125). Hershey, PA: IGI Global. doi:10.4018/978-1-4666-9899-4.ch006

Başal, B. (2017). Actor Effect: A Study on Historical Figures Who Have Shaped the Advertising Narration. In R. Yılmaz (Ed.), *Narrative Advertising Models and Conceptualization in the Digital Age* (pp. 34–60). Hershey, PA: IGI Global. doi:10.4018/978-1-5225-2373-4.ch003

Behjati, M., & Cosmas, J. (2017). Self-Organizing Network Solutions: A Principal Step Towards Real 4G and Beyond. In D. Singh (Ed.), *Routing Protocols and Architectural Solutions for Optimal Wireless Networks and Security* (pp. 241–253). Hershey, PA: IGI Global. doi:10.4018/978-1-5225-2342-0.ch011

Bekafigo, M., & Pingley, A. C. (2017). Do Campaigns "Go Negative" on Twitter? In Y. Ibrahim (Ed.), *Politics, Protest, and Empowerment in Digital Spaces* (pp. 178–191). Hershey, PA: IGI Global. doi:10.4018/978-1-5225-1862-4.ch011

Bender, S., & Dickenson, P. (2016). Utilizing Social Media to Engage Students in Online Learning: Building Relationships Outside of the Learning Management System. In P. Dickenson & J. Jaurez (Eds.), *Increasing Productivity and Efficiency in Online Teaching* (pp. 84–105). Hershey, PA: IGI Global. doi:10.4018/978-1-5225-0347-7.ch005

Bermingham, N., & Prendergast, M. (2016). Bespoke Mobile Application Development: Facilitating Transition of Foundation Students to Higher Education. In L. Briz-Ponce, J. Juanes-Méndez, & F. García-Peñalvo (Eds.), *Handbook of Research on Mobile Devices and Applications in Higher Education Settings* (pp. 222–249). Hershey, PA: IGI Global. doi:10.4018/978-1-5225-0256-2.ch010

Bishop, J. (2017). Developing and Validating the "This Is Why We Can't Have Nice Things Scale": Optimising Political Online Communities for Internet Trolling. In Y. Ibrahim (Ed.), *Politics, Protest, and Empowerment in Digital Spaces* (pp. 153–177). Hershey, PA: IGI Global. doi:10.4018/978-1-5225-1862-4.ch010

Bolat, N. (2017). The Functions of the Narrator in Digital Advertising. In R. Yılmaz (Ed.), *Narrative Advertising Models and Conceptualization in the Digital Age* (pp. 184–201). Hershey, PA: IGI Global. doi:10.4018/978-1-5225-2373-4.ch011

Bowen, G., & Bowen, D. (2016). Social Media: Strategic Decision Making Tool. In W. Ozuem & G. Bowen (Eds.), *Competitive Social Media Marketing Strategies* (pp. 94–111). Hershey, PA: IGI Global. doi:10.4018/978-1-4666-9776-8.ch005

Brown, M. A. Sr. (2017). SNIP: High Touch Approach to Communication. In *Solutions for High-Touch Communications in a High-Tech World* (pp. 71–88). Hershey, PA: IGI Global. doi:10.4018/978-1-5225-1897-6.ch004

Brown, M. A. Sr. (2017). Comparing FTF and Online Communication Knowledge. In *Solutions for High-Touch Communications in a High-Tech World* (pp. 103–113). Hershey, PA: IGI Global. doi:10.4018/978-1-5225-1897-6.ch006

Brown, M. A. Sr. (2017). Where Do We Go from Here? In *Solutions for High-Touch Communications in a High-Tech World* (pp. 137–159). Hershey, PA: IGI Global. doi:10.4018/978-1-5225-1897-6.ch008

Brown, M. A. Sr. (2017). Bridging the Communication Gap. In *Solutions for High-Touch Communications in a High-Tech World* (pp. 1–22). Hershey, PA: IGI Global. doi:10.4018/978-1-5225-1897-6.ch001

Brown, M. A. Sr. (2017). Key Strategies for Communication. In *Solutions for High-Touch Communications in a High-Tech World* (pp. 179–202). Hershey, PA: IGI Global. doi:10.4018/978-1-5225-1897-6.ch010

Bryant, K. N. (2017). WordUp!: Student Responses to Social Media in the Technical Writing Classroom. In K. Bryant (Ed.), *Engaging 21st Century Writers with Social Media* (pp. 231–245). Hershey, PA: IGI Global. doi:10.4018/978-1-5225-0562-4.ch014

Buck, E. H. (2017). Slacktivism, Supervision, and #Selfies: Illuminating Social Media Composition through Reception Theory. In K. Bryant (Ed.), *Engaging 21st Century Writers with Social Media* (pp. 163–178). Hershey, PA: IGI Global. doi:10.4018/978-1-5225-0562-4.ch010

Bucur, B. (2016). Sociological School of Bucharest's Publications and the Romanian Political Propaganda in the Interwar Period. In A. Fox (Ed.), *Global Perspectives on Media Events in Contemporary Society* (pp. 106–120). Hershey, PA: IGI Global. doi:10.4018/978-1-4666-9967-0.ch008

Bull, R., & Pianosi, M. (2017). Social Media, Participation, and Citizenship: New Strategic Directions. In V. Benson, R. Tuninga, & G. Saridakis (Eds.), *Analyzing the Strategic Role of Social Networking in Firm Growth and Productivity* (pp. 76–94). Hershey, PA: IGI Global. doi:10.4018/978-1-5225-0559-4.ch005

Camillo, A. A., & Camillo, I. C. (2016). The Ethics of Strategic Managerial Communication in the Global Context. In A. Normore, L. Long, & M. Javidi (Eds.), *Handbook of Research on Effective Communication, Leadership, and Conflict Resolution* (pp. 566–590). Hershey, PA: IGI Global. doi:10.4018/978-1-4666-9970-0.ch030

Cassard, A., & Sloboda, B. W. (2016). Faculty Perception of Virtual 3-D Learning Environment to Assess Student Learning. In D. Choi, A. Dailey-Hebert, & J. Simmons Estes (Eds.), *Emerging Tools and Applications of Virtual Reality in Education* (pp. 48–74). Hershey, PA: IGI Global. doi:10.4018/978-1-4666-9837-6.ch003

Castellano, S., & Khelladi, I. (2017). Play It Like Beckham!: The Influence of Social Networks on E-Reputation – The Case of Sportspeople and Their Online Fan Base. In A. Mesquita (Ed.), *Research Paradigms and Contemporary Perspectives on Human-Technology Interaction* (pp. 43–61). Hershey, PA: IGI Global. doi:10.4018/978-1-5225-1868-6.ch003

Castellet, A. (2016). What If Devices Take Command: Content Innovation Perspectives for Smart Wearables in the Mobile Ecosystem. *International Journal of Handheld Computing Research*, 7(2), 16–33. doi:10.4018/IJHCR.2016040102

Chugh, R., & Joshi, M. (2017). Challenges of Knowledge Management amidst Rapidly Evolving Tools of Social Media. In R. Chugh (Ed.), *Harnessing Social Media as a Knowledge Management Tool* (pp. 299–314). Hershey, PA: IGI Global. doi:10.4018/978-1-5225-0495-5.ch014

Cockburn, T., & Smith, P. A. (2016). Leadership in the Digital Age: Rhythms and the Beat of Change. In A. Normore, L. Long, & M. Javidi (Eds.), *Handbook of Research on Effective Communication, Leadership, and Conflict Resolution* (pp. 1–20). Hershey, PA: IGI Global. doi:10.4018/978-1-4666-9970-0.ch001

Cole, A. W., & Salek, T. A. (2017). Adopting a Parasocial Connection to Overcome Professional Kakoethos in Online Health Information. In M. Folk & S. Apostel (Eds.), *Establishing and Evaluating Digital Ethos and Online Credibility* (pp. 104–120). Hershey, PA: IGI Global. doi:10.4018/978-1-5225-1072-7.ch006

Cossiavelou, V. (2017). ACTA as Media Gatekeeping Factor: The EU Role as Global Negotiator. *International Journal of Interdisciplinary Telecommunications and Networking*, *9*(1), 26–37. doi:10.4018/IJITN.2017010103

Costanza, F. (2017). Social Media Marketing and Value Co-Creation: A System Dynamics Approach. In S. Rozenes & Y. Cohen (Eds.), *Handbook of Research on Strategic Alliances and Value Co-Creation in the Service Industry* (pp. 205–230). Hershey, PA: IGI Global. doi:10.4018/978-1-5225-2084-9.ch011

Cross, D. E. (2016). Globalization and Media's Impact on Cross Cultural Communication: Managing Organizational Change. In A. Normore, L. Long, & M. Javidi (Eds.), *Handbook of Research on Effective Communication, Leadership, and Conflict Resolution* (pp. 21–41). Hershey, PA: IGI Global. doi:10.4018/978-1-4666-9970-0.ch002

Damásio, M. J., Henriques, S., Teixeira-Botelho, I., & Dias, P. (2016). Mobile Media and Social Interaction: Mobile Services and Content as Drivers of Social Interaction. In J. Aguado, C. Feijóo, & I. Martínez (Eds.), *Emerging Perspectives on the Mobile Content Evolution* (pp. 357–379). Hershey, PA: IGI Global. doi:10.4018/978-1-4666-8838-4.ch018

Davis, A., & Foley, L. (2016). Digital Storytelling. In B. Guzzetti & M. Lesley (Eds.), *Handbook of Research on the Societal Impact of Digital Media* (pp. 317–342). Hershey, PA: IGI Global. doi:10.4018/978-1-4666-8310-5.ch013

Davis, S., Palmer, L., & Etienne, J. (2016). The Geography of Digital Literacy: Mapping Communications Technology Training Programs in Austin, Texas. In B. Passarelli, J. Straubhaar, & A. Cuevas-Cerveró (Eds.), *Handbook of Research on Comparative Approaches to the Digital Age Revolution in Europe and the Americas* (pp. 371–384). Hershey, PA: IGI Global. doi:10.4018/978-1-4666-8740-0.ch022

Delello, J. A., & McWhorter, R. R. (2016). New Visual Literacies and Competencies for Education and the Workplace. In B. Guzzetti & M. Lesley (Eds.), *Handbook of Research on the Societal Impact of Digital Media* (pp. 127–162). Hershey, PA: IGI Global. doi:10.4018/978-1-4666-8310-5.ch006

Di Virgilio, F., & Antonelli, G. (2018). Consumer Behavior, Trust, and Electronic Word-of-Mouth Communication: Developing an Online Purchase Intention Model. In F. Di Virgilio (Ed.), *Social Media for Knowledge Management Applications in Modern Organizations* (pp. 58–80). Hershey, PA: IGI Global. doi:10.4018/978-1-5225-2897-5.ch003

Dixit, S. K. (2016). eWOM Marketing in Hospitality Industry. In A. Singh, & P. Duhan (Eds.), Managing Public Relations and Brand Image through Social Media (pp. 266-280). Hershey, PA: IGI Global. doi:10.4018/978-1-5225-0332-3.ch014

Duhan, P., & Singh, A. (2016). Facebook Experience Is Different: An Empirical Study in Indian Context. In S. Rathore & A. Panwar (Eds.), *Capturing, Analyzing, and Managing Word-of-Mouth in the Digital Marketplace* (pp. 188–212). Hershey, PA: IGI Global. doi:10.4018/978-1-4666-9449-1.ch011

Dunne, D. J. (2016). The Scholar's Ludo-Narrative Game and Multimodal Graphic Novel: A Comparison of Fringe Scholarship. In A. Connor & S. Marks (Eds.), *Creative Technologies for Multidisciplinary Applications* (pp. 182–207). Hershey, PA: IGI Global. doi:10.4018/978-1-5225-0016-2.ch008

DuQuette, J. L. (2017). Lessons from Cypris Chat: Revisiting Virtual Communities as Communities. In G. Panconesi & M. Guida (Eds.), *Handbook of Research on Collaborative Teaching Practice in Virtual Learning Environments* (pp. 299–316). Hershey, PA: IGI Global. doi:10.4018/978-1-5225-2426-7.ch016

Ekhlassi, A., Niknejhad Moghadam, M., & Adibi, A. (2018). The Concept of Social Media: The Functional Building Blocks. In *Building Brand Identity in the Age of Social Media: Emerging Research and Opportunities* (pp. 29–60). Hershey, PA: IGI Global. doi:10.4018/978-1-5225-5143-0.ch002

Ekhlassi, A., Niknejhad Moghadam, M., & Adibi, A. (2018). Social Media Branding Strategy: Social Media Marketing Approach. In *Building Brand Identity in the Age of Social Media: Emerging Research and Opportunities* (pp. 94–117). Hershey, PA: IGI Global. doi:10.4018/978-1-5225-5143-0.ch004

Ekhlassi, A., Niknejhad Moghadam, M., & Adibi, A. (2018). The Impact of Social Media on Brand Loyalty: Achieving "E-Trust" Through Engagement. In *Building Brand Identity in the Age of Social Media: Emerging Research and Opportunities* (pp. 155–168). Hershey, PA: IGI Global. doi:10.4018/978-1-5225-5143-0.ch007

Elegbe, O. (2017). An Assessment of Media Contribution to Behaviour Change and HIV Prevention in Nigeria. In O. Nelson, B. Ojebuyi, & A. Salawu (Eds.), *Impacts of the Media on African Socio-Economic Development* (pp. 261–280). Hershey, PA: IGI Global. doi:10.4018/978-1-5225-1859-4.ch017

Endong, F. P. (2018). Hashtag Activism and the Transnationalization of Nigerian-Born Movements Against Terrorism: A Critical Appraisal of the #BringBackOurGirls Campaign. In F. Endong (Ed.), *Exploring the Role of Social Media in Transnational Advocacy* (pp. 36–54). Hershey, PA: IGI Global. doi:10.4018/978-1-5225-2854-8.ch003

Erragcha, N. (2017). Using Social Media Tools in Marketing: Opportunities and Challenges. In M. Brown Sr., (Ed.), *Social Media Performance Evaluation and Success Measurements* (pp. 106–129). Hershey, PA: IGI Global. doi:10.4018/978-1-5225-1963-8.ch006

Ezeh, N. C. (2018). Media Campaign on Exclusive Breastfeeding: Awareness, Perception, and Acceptability Among Mothers in Anambra State, Nigeria. In A. Salawu & T. Owolabi (Eds.), *Exploring Journalism Practice and Perception in Developing Countries* (pp. 172–193). Hershey, PA: IGI Global. doi:10.4018/978-1-5225-3376-4.ch009

Fawole, O. A., & Osho, O. A. (2017). Influence of Social Media on Dating Relationships of Emerging Adults in Nigerian Universities: Social Media and Dating in Nigeria. In M. Wright (Ed.), *Identity, Sexuality, and Relationships among Emerging Adults in the Digital Age* (pp. 168–177). Hershey, PA: IGI Global. doi:10.4018/978-1-5225-1856-3.ch011

Fayoyin, A. (2017). Electoral Polling and Reporting in Africa: Professional and Policy Implications for Media Practice and Political Communication in a Digital Age. In N. Mhiripiri & T. Chari (Eds.), *Media Law, Ethics, and Policy in the Digital Age* (pp. 164–181). Hershey, PA: IGI Global. doi:10.4018/978-1-5225-2095-5.ch009

Fayoyin, A. (2018). Rethinking Media Engagement Strategies for Social Change in Africa: Context, Approaches, and Implications for Development Communication. In A. Salawu & T. Owolabi (Eds.), *Exploring Journalism Practice and Perception in Developing Countries* (pp. 257–280). Hershey, PA: IGI Global. doi:10.4018/978-1-5225-3376-4.ch013

Fechine, Y., & Rêgo, S. C. (2018). Transmedia Television Journalism in Brazil: Jornal da Record News as Reference. In R. Gambarato & G. Alzamora (Eds.), *Exploring Transmedia Journalism in the Digital Age* (pp. 253–265). Hershey, PA: IGI Global. doi:10.4018/978-1-5225-3781-6.ch015

Feng, J., & Lo, K. (2016). Video Broadcasting Protocol for Streaming Applications with Cooperative Clients. In D. Kanellopoulos (Ed.), *Emerging Research on Networked Multimedia Communication Systems* (pp. 205–229). Hershey, PA: IGI Global. doi:10.4018/978-1-4666-8850-6.ch006

Fiore, C. (2017). The Blogging Method: Improving Traditional Student Writing Practices. In K. Bryant (Ed.), *Engaging 21st Century Writers with Social Media* (pp. 179–198). Hershey, PA: IGI Global. doi:10.4018/978-1-5225-0562-4.ch011

Fleming, J., & Kajimoto, M. (2016). The Freedom of Critical Thinking: Examining Efforts to Teach American News Literacy Principles in Hong Kong, Vietnam, and Malaysia. In M. Yildiz & J. Keengwe (Eds.), *Handbook of Research on Media Literacy in the Digital Age* (pp. 208–235). Hershey, PA: IGI Global. doi:10.4018/978-1-4666-9667-9.ch010

Gambarato, R. R., Alzamora, G. C., & Tárcia, L. P. (2018). 2016 Rio Summer Olympics and the Transmedia Journalism of Planned Events. In R. Gambarato & G. Alzamora (Eds.), *Exploring Transmedia Journalism in the Digital Age* (pp. 126–146). Hershey, PA: IGI Global. doi:10.4018/978-1-5225-3781-6.ch008

Ganguin, S., Gemkow, J., & Haubold, R. (2017). Information Overload as a Challenge and Changing Point for Educational Media Literacies. In R. Marques & J. Batista (Eds.), *Information and Communication Overload in the Digital Age* (pp. 302–328). Hershey, PA: IGI Global. doi:10.4018/978-1-5225-2061-0.ch013

Gao, Y. (2016). Reviewing Gratification Effects in Mobile Gaming. In X. Xu (Ed.), *Handbook of Research on Human Social Interaction in the Age of Mobile Devices* (pp. 406–428). Hershey, PA: IGI Global. doi:10.4018/978-1-5225-0469-6.ch017

Gardner, G. C. (2017). The Lived Experience of Smartphone Use in a Unit of the United States Army. In F. Topor (Ed.), *Handbook of Research on Individualism and Identity in the Globalized Digital Age* (pp. 88–117). Hershey, PA: IGI Global. doi:10.4018/978-1-5225-0522-8.ch005

Giessen, H. W. (2016). The Medium, the Content, and the Performance: An Overview on Media-Based Learning. In B. Khan (Ed.), *Revolutionizing Modern Education through Meaningful E-Learning Implementation* (pp. 42–55). Hershey, PA: IGI Global. doi:10.4018/978-1-5225-0466-5.ch003

Giltenane, J. (2016). Investigating the Intention to Use Social Media Tools Within Virtual Project Teams. In G. Silvius (Ed.), *Strategic Integration of Social Media into Project Management Practice* (pp. 83–105). Hershey, PA: IGI Global. doi:10.4018/978-1-4666-9867-3.ch006

Golightly, D., & Houghton, R. J. (2018). Social Media as a Tool to Understand Behaviour on the Railways. In S. Kohli, A. Kumar, J. Easton, & C. Roberts (Eds.), *Innovative Applications of Big Data in the Railway Industry* (pp. 224–239). Hershey, PA: IGI Global. doi:10.4018/978-1-5225-3176-0.ch010

Goovaerts, M., Nieuwenhuysen, P., & Dhamdhere, S. N. (2016). VLIR-UOS Workshop 'E-Info Discovery and Management for Institutes in the South': Presentations and Conclusions, Antwerp, 8-19 December, 2014. In E. de Smet, & S. Dhamdhere (Eds.), E-Discovery Tools and Applications in Modern Libraries (pp. 1-40). Hershey, PA: IGI Global. doi:10.4018/978-1-5225-0474-0.ch001

Grützmann, A., Carvalho de Castro, C., Meireles, A. A., & Rodrigues, R. C. (2016). Organizational Architecture and Online Social Networks: Insights from Innovative Brazilian Companies. In G. Jamil, J. Poças Rascão, F. Ribeiro, & A. Malheiro da Silva (Eds.), *Handbook of Research on Information Architecture and Management in Modern Organizations* (pp. 508–524). Hershey, PA: IGI Global. doi:10.4018/978-1-4666-8637-3.ch023

Gundogan, M. B. (2017). In Search for a "Good Fit" Between Augmented Reality and Mobile Learning Ecosystem. In G. Kurubacak & H. Altinpulluk (Eds.), *Mobile Technologies and Augmented Reality in Open Education* (pp. 135–153). Hershey, PA: IGI Global. doi:10.4018/978-1-5225-2110-5.ch007

Gupta, H. (2018). Impact of Digital Communication on Consumer Behaviour Processes in Luxury Branding Segment: A Study of Apparel Industry. In S. Dasgupta, S. Biswal, & M. Ramesh (Eds.), *Holistic Approaches to Brand Culture and Communication Across Industries* (pp. 132–157). Hershey, PA: IGI Global. doi:10.4018/978-1-5225-3150-0.ch008

Hai-Jew, S. (2017). Creating "(Social) Network Art" with NodeXL. In S. Hai-Jew (Ed.), *Social Media Data Extraction and Content Analysis* (pp. 342–393). Hershey, PA: IGI Global. doi:10.4018/978-1-5225-0648-5.ch011

Hai-Jew, S. (2017). Employing the Sentiment Analysis Tool in NVivo 11 Plus on Social Media Data: Eight Initial Case Types. In N. Rao (Ed.), *Social Media Listening and Monitoring for Business Applications* (pp. 175–244). Hershey, PA: IGI Global. doi:10.4018/978-1-5225-0846-5.ch010

Hai-Jew, S. (2017). Conducting Sentiment Analysis and Post-Sentiment Data Exploration through Automated Means. In S. Hai-Jew (Ed.), *Social Media Data Extraction and Content Analysis* (pp. 202–240). Hershey, PA: IGI Global. doi:10.4018/978-1-5225-0648-5.ch008

Hai-Jew, S. (2017). Applied Analytical "Distant Reading" using NVivo 11 Plus. In S. Hai-Jew (Ed.), *Social Media Data Extraction and Content Analysis* (pp. 159–201). Hershey, PA: IGI Global. doi:10.4018/978-1-5225-0648-5.ch007

Hai-Jew, S. (2017). Flickering Emotions: Feeling-Based Associations from Related Tags Networks on Flickr. In S. Hai-Jew (Ed.), *Social Media Data Extraction and Content Analysis* (pp. 296–341). Hershey, PA: IGI Global. doi:10.4018/978-1-5225-0648-5.ch010

Hai-Jew, S. (2017). Manually Profiling Egos and Entities across Social Media Platforms: Evaluating Shared Messaging and Contents, User Networks, and Metadata. In V. Benson, R. Tuninga, & G. Saridakis (Eds.), *Analyzing the Strategic Role of Social Networking in Firm Growth and Productivity* (pp. 352–405). Hershey, PA: IGI Global. doi:10.4018/978-1-5225-0559-4.ch019

Hai-Jew, S. (2017). Exploring "User," "Video," and (Pseudo) Multi-Mode Networks on YouTube with NodeXL. In S. Hai-Jew (Ed.), *Social Media Data Extraction and Content Analysis* (pp. 242–295). Hershey, PA: IGI Global. doi:10.4018/978-1-5225-0648-5.ch009

Hai-Jew, S. (2018). Exploring "Mass Surveillance" Through Computational Linguistic Analysis of Five Text Corpora: Academic, Mainstream Journalism, Microblogging Hashtag Conversation, Wikipedia Articles, and Leaked Government Data. In *Techniques for Coding Imagery and Multimedia: Emerging Research and Opportunities* (pp. 212–286). Hershey, PA: IGI Global. doi:10.4018/978-1-5225-2679-7.ch004

Hai-Jew, S. (2018). Exploring Identity-Based Humor in a #Selfies #Humor Image Set From Instagram. In *Techniques for Coding Imagery and Multimedia: Emerging Research and Opportunities* (pp. 1–90). Hershey, PA: IGI Global. doi:10.4018/978-1-5225-2679-7.ch001

Hai-Jew, S. (2018). See Ya!: Exploring American Renunciation of Citizenship Through Targeted and Sparse Social Media Data Sets and a Custom Spatial-Based Linguistic Analysis Dictionary. In *Techniques for Coding Imagery and Multimedia: Emerging Research and Opportunities* (pp. 287–393). Hershey, PA: IGI Global. doi:10.4018/978-1-5225-2679-7.ch005

Han, H. S., Zhang, J., Peikazadi, N., Shi, G., Hung, A., Doan, C. P., & Filippelli, S. (2016). An Entertaining Game-Like Learning Environment in a Virtual World for Education. In S. D'Agustino (Ed.), *Creating Teacher Immediacy in Online Learning Environments* (pp. 290–306). Hershey, PA: IGI Global. doi:10.4018/978-1-4666-9995-3.ch015

Harrin, E. (2016). Barriers to Social Media Adoption on Projects. In G. Silvius (Ed.), *Strategic Integration of Social Media into Project Management Practice* (pp. 106–124). Hershey, PA: IGI Global. doi:10.4018/978-1-4666-9867-3.ch007

Harvey, K. E. (2016). Local News and Mobile: Major Tipping Points. In X. Xu (Ed.), *Handbook of Research on Human Social Interaction in the Age of Mobile Devices* (pp. 171–199). Hershey, PA: IGI Global. doi:10.4018/978-1-5225-0469-6.ch009

Harvey, K. E., & An, Y. (2016). Marketing and Mobile: Increasing Integration. In X. Xu (Ed.), *Handbook of Research on Human Social Interaction in the Age of Mobile Devices* (pp. 220–247). Hershey, PA: IGI Global. doi:10.4018/978-1-5225-0469-6.ch011

Harvey, K. E., Auter, P. J., & Stevens, S. (2016). Educators and Mobile: Challenges and Trends. In X. Xu (Ed.), *Handbook of Research on Human Social Interaction in the Age of Mobile Devices* (pp. 61–95). Hershey, PA: IGI Global. doi:10.4018/978-1-5225-0469-6.ch004

Hasan, H., & Linger, H. (2017). Connected Living for Positive Ageing. In S. Gordon (Ed.), *Online Communities as Agents of Change and Social Movements* (pp. 203–223). Hershey, PA: IGI Global. doi:10.4018/978-1-5225-2495-3.ch008

Hashim, K., Al-Sharqi, L., & Kutbi, I. (2016). Perceptions of Social Media Impact on Social Behavior of Students: A Comparison between Students and Faculty. *International Journal of Virtual Communities and Social Networking*, 8(2), 1–11. doi:10.4018/IJVCSN.2016040101

Henriques, S., & Damasio, M. J. (2016). The Value of Mobile Communication for Social Belonging: Mobile Apps and the Impact on Social Interaction. *International Journal of Handheld Computing Research*, 7(2), 44–58. doi:10.4018/IJHCR.2016040104

Hersey, L. N. (2017). CHOICES: Measuring Return on Investment in a Nonprofit Organization. In M. Brown Sr., (Ed.), *Social Media Performance Evaluation and Success Measurements* (pp. 157–179). Hershey, PA: IGI Global. doi:10.4018/978-1-5225-1963-8.ch008

Heuva, W. E. (2017). Deferring Citizens' "Right to Know" in an Information Age: The Information Deficit in Namibia. In N. Mhiripiri & T. Chari (Eds.), *Media Law, Ethics, and Policy in the Digital Age* (pp. 245–267). Hershey, PA: IGI Global. doi:10.4018/978-1-5225-2095-5.ch014

Hopwood, M., & McLean, H. (2017). Social Media in Crisis Communication: The Lance Armstrong Saga. In V. Benson, R. Tuninga, & G. Saridakis (Eds.), *Analyzing the Strategic Role of Social Networking in Firm Growth and Productivity* (pp. 45–58). Hershey, PA: IGI Global. doi:10.4018/978-1-5225-0559-4.ch003

Hotur, S. K. (2018). Indian Approaches to E-Diplomacy: An Overview. In S. Bute (Ed.), *Media Diplomacy and Its Evolving Role in the Current Geopolitical Climate* (pp. 27–35). Hershey, PA: IGI Global. doi:10.4018/978-1-5225-3859-2.ch002

Ibadildin, N., & Harvey, K. E. (2016). Business and Mobile: Rapid Restructure Required. In X. Xu (Ed.), *Handbook of Research on Human Social Interaction in the Age of Mobile Devices* (pp. 312–350). Hershey, PA: IGI Global. doi:10.4018/978-1-5225-0469-6.ch014

Iwasaki, Y. (2017). Youth Engagement in the Era of New Media. In M. Adria & Y. Mao (Eds.), *Handbook of Research on Citizen Engagement and Public Participation in the Era of New Media* (pp. 90–105). Hershey, PA: IGI Global. doi:10.4018/978-1-5225-1081-9.ch006

Jamieson, H. V. (2017). We have a Situation!: Cyberformance and Civic Engagement in Post-Democracy. In R. Shin (Ed.), *Convergence of Contemporary Art, Visual Culture, and Global Civic Engagement* (pp. 297–317). Hershey, PA: IGI Global. doi:10.4018/978-1-5225-1665-1.ch017

Jimoh, J., & Kayode, J. (2018). Imperative of Peace and Conflict-Sensitive Journalism in Development. In A. Salawu & T. Owolabi (Eds.), *Exploring Journalism Practice and Perception in Developing Countries* (pp. 150–171). Hershey, PA: IGI Global. doi:10.4018/978-1-5225-3376-4.ch008

Johns, R. (2016). Increasing Value of a Tangible Product through Intangible Attributes: Value Co-Creation and Brand Building within Online Communities – Virtual Communities and Value. In R. English & R. Johns (Eds.), *Gender Considerations in Online Consumption Behavior and Internet Use* (pp. 112–124). Hershey, PA: IGI Global. doi:10.4018/978-1-5225-0010-0.ch008

Kanellopoulos, D. N. (2018). Group Synchronization for Multimedia Systems. In M. Khosrow-Pour, D.B.A. (Ed.), Encyclopedia of Information Science and Technology, Fourth Edition (pp. 6435-6446). Hershey, PA: IGI Global. doi:10.4018/978-1-5225-2255-3.ch559

Kapepo, M. I., & Mayisela, T. (2017). Integrating Digital Literacies Into an Undergraduate Course: Inclusiveness Through Use of ICTs. In C. Ayo & V. Mbarika (Eds.), *Sustainable ICT Adoption and Integration for Socio-Economic Development* (pp. 152–173). Hershey, PA: IGI Global. doi:10.4018/978-1-5225-2565-3.ch007

Karahoca, A., & Yengin, İ. (2018). Understanding the Potentials of Social Media in Collaborative Learning. In M. Khosrow-Pour, D.B.A. (Ed.), Encyclopedia of Information Science and Technology, Fourth Edition (pp. 7168-7180). Hershey, PA: IGI Global. doi:10.4018/978-1-5225-2255-3.ch623

Karataş, S., Ceran, O., Ülker, Ü., Gün, E. T., Köse, N. Ö., Kılıç, M., ... Tok, Z. A. (2016). A Trend Analysis of Mobile Learning. In D. Parsons (Ed.), *Mobile and Blended Learning Innovations for Improved Learning Outcomes* (pp. 248–276). Hershey, PA: IGI Global. doi:10.4018/978-1-5225-0359-0.ch013

Kasemsap, K. (2016). Role of Social Media in Brand Promotion: An International Marketing Perspective. In A. Singh & P. Duhan (Eds.), *Managing Public Relations and Brand Image through Social Media* (pp. 62–88). Hershey, PA: IGI Global. doi:10.4018/978-1-5225-0332-3.ch005

Kasemsap, K. (2016). The Roles of Social Media Marketing and Brand Management in Global Marketing. In W. Ozuem & G. Bowen (Eds.), *Competitive Social Media Marketing Strategies* (pp. 173–200). Hershey, PA: IGI Global. doi:10.4018/978-1-4666-9776-8.ch009

Kasemsap, K. (2017). Professional and Business Applications of Social Media Platforms. In V. Benson, R. Tuninga, & G. Saridakis (Eds.), *Analyzing the Strategic Role of Social Networking in Firm Growth and Productivity* (pp. 427–450). Hershey, PA: IGI Global. doi:10.4018/978-1-5225-0559-4.ch021

Kasemsap, K. (2017). Mastering Social Media in the Modern Business World. In N. Rao (Ed.), *Social Media Listening and Monitoring for Business Applications* (pp. 18–44). Hershey, PA: IGI Global. doi:10.4018/978-1-5225-0846-5.ch002

Kato, Y., & Kato, S. (2016). Mobile Phone Use during Class at a Japanese Women's College. In M. Yildiz & J. Keengwe (Eds.), *Handbook of Research on Media Literacy in the Digital Age* (pp. 436–455). Hershey, PA: IGI Global. doi:10.4018/978-1-4666-9667-9.ch021

Kaufmann, H. R., & Manarioti, A. (2017). Consumer Engagement in Social Media Platforms. In *Encouraging Participative Consumerism Through Evolutionary Digital Marketing: Emerging Research and Opportunities* (pp. 95–123). Hershey, PA: IGI Global. doi:10.4018/978-1-68318-012-8.ch004

Kavoura, A., & Kefallonitis, E. (2018). The Effect of Social Media Networking in the Travel Industry. In M. Khosrow-Pour, D.B.A. (Ed.), Encyclopedia of Information Science and Technology, Fourth Edition (pp. 4052-4063). Hershey, PA: IGI Global. doi:10.4018/978-1-5225-2255-3.ch351

Kawamura, Y. (2018). Practice and Modeling of Advertising Communication Strategy: Sender-Driven and Receiver-Driven. In T. Ogata & S. Asakawa (Eds.), *Content Generation Through Narrative Communication and Simulation* (pp. 358–379). Hershey, PA: IGI Global. doi:10.4018/978-1-5225-4775-4.ch013

Kell, C., & Czerniewicz, L. (2017). Visibility of Scholarly Research and Changing Research Communication Practices: A Case Study from Namibia. In A. Esposito (Ed.), *Research 2.0 and the Impact of Digital Technologies on Scholarly Inquiry* (pp. 97–116). Hershey, PA: IGI Global. doi:10.4018/978-1-5225-0830-4.ch006

Khalil, G. E. (2016). Change through Experience: How Experiential Play and Emotional Engagement Drive Health Game Success. In D. Novák, B. Tulu, & H. Brendryen (Eds.), *Handbook of Research on Holistic Perspectives in Gamification for Clinical Practice* (pp. 10–34). Hershey, PA: IGI Global. doi:10.4018/978-1-4666-9522-1.ch002

Kılınç, U. (2017). Create It! Extend It!: Evolution of Comics Through Narrative Advertising. In R. Yılmaz (Ed.), *Narrative Advertising Models and Conceptualization in the Digital Age* (pp. 117–132). Hershey, PA: IGI Global. doi:10.4018/978-1-5225-2373-4.ch007

Kim, J. H. (2016). Pedagogical Approaches to Media Literacy Education in the United States. In M. Yildiz & J. Keengwe (Eds.), *Handbook of Research on Media Literacy in the Digital Age* (pp. 53–74). Hershey, PA: IGI Global. doi:10.4018/978-1-4666-9667-9.ch003

Kirigha, J. M., Mukhongo, L. L., & Masinde, R. (2016). Beyond Web 2.0. Social Media and Urban Educated Youths Participation in Kenyan Politics. In L. Mukhongo & J. Macharia (Eds.), *Political Influence of the Media in Developing Countries* (pp. 156–174). Hershey, PA: IGI Global. doi:10.4018/978-1-4666-9613-6.ch010

Krochmal, M. M. (2016). Training for Mobile Journalism. In D. Mentor (Ed.), *Handbook of Research on Mobile Learning in Contemporary Classrooms* (pp. 336–362). Hershey, PA: IGI Global. doi:10.4018/978-1-5225-0251-7.ch017

Kumar, P., & Sinha, A. (2018). Business-Oriented Analytics With Social Network of Things. In H. Bansal, G. Shrivastava, G. Nguyen, & L. Stanciu (Eds.), *Social Network Analytics for Contemporary Business Organizations* (pp. 166–187). Hershey, PA: IGI Global. doi:10.4018/978-1-5225-5097-6.ch009

Kunock, A. I. (2017). Boko Haram Insurgency in Cameroon: Role of Mass Media in Conflict Management. In N. Mhiripiri & T. Chari (Eds.), *Media Law, Ethics, and Policy in the Digital Age* (pp. 226–244). Hershey, PA: IGI Global. doi:10.4018/978-1-5225-2095-5.ch013

Labadie, J. A. (2018). Digitally Mediated Art Inspired by Technology Integration: A Personal Journey. In A. Ursyn (Ed.), *Visual Approaches to Cognitive Education With Technology Integration* (pp. 121–162). Hershey, PA: IGI Global. doi:10.4018/978-1-5225-5332-8.ch008

Lefkowith, S. (2017). Credibility and Crisis in Pseudonymous Communities. In M. Folk & S. Apostel (Eds.), *Establishing and Evaluating Digital Ethos and Online Credibility* (pp. 190–236). Hershey, PA: IGI Global. doi:10.4018/978-1-5225-1072-7.ch010

Lemoine, P. A., Hackett, P. T., & Richardson, M. D. (2016). The Impact of Social Media on Instruction in Higher Education. In L. Briz-Ponce, J. Juanes-Méndez, & F. García-Peñalvo (Eds.), *Handbook of Research on Mobile Devices and Applications in Higher Education Settings* (pp. 373–401). Hershey, PA: IGI Global. doi:10.4018/978-1-5225-0256-2.ch016

Liampotis, N., Papadopoulou, E., Kalatzis, N., Roussaki, I. G., Kosmides, P., Sykas, E. D., ... Taylor, N. K. (2016). Tailoring Privacy-Aware Trustworthy Cooperating Smart Spaces for University Environments. In A. Panagopoulos (Ed.), *Handbook of Research on Next Generation Mobile Communication Systems* (pp. 410–439). Hershey, PA: IGI Global. doi:10.4018/978-1-4666-8732-5.ch016

Luppicini, R. (2017). Technoethics and Digital Democracy for Future Citizens. In R. Luppicini & R. Baarda (Eds.), *Digital Media Integration for Participatory Democracy* (pp. 1–21). Hershey, PA: IGI Global. doi:10.4018/978-1-5225-2463-2.ch001

Mahajan, I. M., Rather, M., Shafiq, H., & Qadri, U. (2016). Media Literacy Organizations. In M. Yildiz & J. Keengwe (Eds.), *Handbook of Research on Media Literacy in the Digital Age* (pp. 236–248). Hershey, PA: IGI Global. doi:10.4018/978-1-4666-9667-9.ch011

Maher, D. (2018). Supporting Pre-Service Teachers' Understanding and Use of Mobile Devices. In J. Keengwe (Ed.), *Handbook of Research on Mobile Technology, Constructivism, and Meaningful Learning* (pp. 160–177). Hershey, PA: IGI Global. doi:10.4018/978-1-5225-3949-0.ch009

Makhwanya, A. (2018). Barriers to Social Media Advocacy: Lessons Learnt From the Project "Tell Them We Are From Here". In F. Endong (Ed.), *Exploring the Role of Social Media in Transnational Advocacy* (pp. 55–72). Hershey, PA: IGI Global. doi:10.4018/978-1-5225-2854-8.ch004

Manli, G., & Rezaei, S. (2017). Value and Risk: Dual Pillars of Apps Usefulness. In S. Rezaei (Ed.), *Apps Management and E-Commerce Transactions in Real-Time* (pp. 274–292). Hershey, PA: IGI Global. doi:10.4018/978-1-5225-2449-6.ch013

Manrique, C. G., & Manrique, G. G. (2017). Social Media's Role in Alleviating Political Corruption and Scandals: The Philippines during and after the Marcos Regime. In K. Demirhan & D. Çakır-Demirhan (Eds.), *Political Scandal, Corruption, and Legitimacy in the Age of Social Media* (pp. 205–222). Hershey, PA: IGI Global. doi:10.4018/978-1-5225-2019-1.ch009

Manzoor, A. (2016). Cultural Barriers to Organizational Social Media Adoption. In A. Goel & P. Singhal (Eds.), *Product Innovation through Knowledge Management and Social Media Strategies* (pp. 31–45). Hershey, PA: IGI Global. doi:10.4018/978-1-4666-9607-5.ch002

Manzoor, A. (2016). Social Media for Project Management. In G. Silvius (Ed.), *Strategic Integration of Social Media into Project Management Practice* (pp. 51–65). Hershey, PA: IGI Global. doi:10.4018/978-1-4666-9867-3.ch004

Marovitz, M. (2017). Social Networking Engagement and Crisis Communication Considerations. In M. Brown Sr., (Ed.), *Social Media Performance Evaluation and Success Measurements* (pp. 130–155). Hershey, PA: IGI Global. doi:10.4018/978-1-5225-1963-8.ch007

Mathur, D., & Mathur, D. (2016). Word of Mouth on Social Media: A Potent Tool for Brand Building. In S. Rathore & A. Panwar (Eds.), *Capturing, Analyzing, and Managing Word-of-Mouth in the Digital Marketplace* (pp. 45–60). Hershey, PA: IGI Global. doi:10.4018/978-1-4666-9449-1.ch003

Maulana, I. (2018). Spontaneous Taking and Posting Selfie: Reclaiming the Lost Trust. In S. Hai-Jew (Ed.), *Selfies as a Mode of Social Media and Work Space Research* (pp. 28–50). Hershey, PA: IGI Global. doi:10.4018/978-1-5225-3373-3.ch002

Mayo, S. (2018). A Collective Consciousness Model in a Post-Media Society. In M. Khosrow-Pour (Ed.), *Enhancing Art, Culture, and Design With Technological Integration* (pp. 25–49). Hershey, PA: IGI Global. doi:10.4018/978-1-5225-5023-5.ch002

Mazur, E., Signorella, M. L., & Hough, M. (2018). The Internet Behavior of Older Adults. In M. Khosrow-Pour, D.B.A. (Ed.), Encyclopedia of Information Science and Technology, Fourth Edition (pp. 7026-7035). Hershey, PA: IGI Global. doi:10.4018/978-1-5225-2255-3.ch609

McGuire, M. (2017). Reblogging as Writing: The Role of Tumblr in the Writing Classroom. In K. Bryant (Ed.), *Engaging 21st Century Writers with Social Media* (pp. 116–131). Hershey, PA: IGI Global. doi:10.4018/978-1-5225-0562-4.ch007

McKee, J. (2018). Architecture as a Tool to Solve Business Planning Problems. In M. Khosrow-Pour, D.B.A. (Ed.), Encyclopedia of Information Science and Technology, Fourth Edition (pp. 573-586). Hershey, PA: IGI Global. doi:10.4018/978-1-5225-2255-3.ch050

McMahon, D. (2017). With a Little Help from My Friends: The Irish Radio Industry's Strategic Appropriation of Facebook for Commercial Growth. In V. Benson, R. Tuninga, & G. Saridakis (Eds.), *Analyzing the Strategic Role of Social Networking in Firm Growth and Productivity* (pp. 157–171). Hershey, PA: IGI Global. doi:10.4018/978-1-5225-0559-4.ch009

McPherson, M. J., & Lemon, N. (2017). The Hook, Woo, and Spin: Academics Creating Relations on Social Media. In A. Esposito (Ed.), *Research 2.0 and the Impact of Digital Technologies on Scholarly Inquiry* (pp. 167–187). Hershey, PA: IGI Global. doi:10.4018/978-1-5225-0830-4.ch009

Melro, A., & Oliveira, L. (2018). Screen Culture. In M. Khosrow-Pour, D.B.A. (Ed.), Encyclopedia of Information Science and Technology, Fourth Edition (pp. 4255-4266). Hershey, PA: IGI Global. doi:10.4018/978-1-5225-2255-3.ch369

Merwin, G. A. Jr, McDonald, J. S., Bennett, J. R. Jr, & Merwin, K. A. (2016). Social Media Applications Promote Constituent Involvement in Government Management. In G. Silvius (Ed.), *Strategic Integration of Social Media into Project Management Practice* (pp. 272–291). Hershey, PA: IGI Global. doi:10.4018/978-1-4666-9867-3.ch016

Mhiripiri, N. A., & Chikakano, J. (2017). Criminal Defamation, the Criminalisation of Expression, Media and Information Dissemination in the Digital Age: A Legal and Ethical Perspective. In N. Mhiripiri & T. Chari (Eds.), *Media Law, Ethics, and Policy in the Digital Age* (pp. 1–24). Hershey, PA: IGI Global. doi:10.4018/978-1-5225-2095-5.ch001

Miliopoulou, G., & Cossiavelou, V. (2016). Brands and Media Gatekeeping in the Social Media: Current Trends and Practices – An Exploratory Research. *International Journal of Interdisciplinary Telecommunications and Networking*, 8(4), 51–64. doi:10.4018/IJITN.2016100105

Miron, E., Palmor, A., Ravid, G., Sharon, A., Tikotsky, A., & Zirkel, Y. (2017). Principles and Good Practices for Using Wikis within Organizations. In R. Chugh (Ed.), *Harnessing Social Media as a Knowledge Management Tool* (pp. 143–176). Hershey, PA: IGI Global. doi:10.4018/978-1-5225-0495-5.ch008

Mishra, K. E., Mishra, A. K., & Walker, K. (2016). Leadership Communication, Internal Marketing, and Employee Engagement: A Recipe to Create Brand Ambassadors. In A. Normore, L. Long, & M. Javidi (Eds.), *Handbook of Research on Effective Communication, Leadership, and Conflict Resolution* (pp. 311–329). Hershey, PA: IGI Global. doi:10.4018/978-1-4666-9970-0.ch017

Moeller, C. L. (2018). Sharing Your Personal Medical Experience Online: Is It an Irresponsible Act or Patient Empowerment? In S. Sekalala & B. Niezgoda (Eds.), *Global Perspectives on Health Communication in the Age of Social Media* (pp. 185–209). Hershey, PA: IGI Global. doi:10.4018/978-1-5225-3716-8.ch007

Mosanako, S. (2017). Broadcasting Policy in Botswana: The Case of Botswana Television. In O. Nelson, B. Ojebuyi, & A. Salawu (Eds.), *Impacts of the Media on African Socio-Economic Development* (pp. 217–230). Hershey, PA: IGI Global. doi:10.4018/978-1-5225-1859-4.ch014

Nazari, A. (2016). Developing a Social Media Communication Plan. In G. Silvius (Ed.), *Strategic Integration of Social Media into Project Management Practice* (pp. 194–217). Hershey, PA: IGI Global. doi:10.4018/978-1-4666-9867-3.ch012

Neto, B. M. (2016). From Information Society to Community Service: The Birth of E-Citizenship. In B. Passarelli, J. Straubhaar, & A. Cuevas-Cerveró (Eds.), *Handbook of Research on Comparative Approaches to the Digital Age Revolution in Europe and the Americas* (pp. 101–123). Hershey, PA: IGI Global. doi:10.4018/978-1-4666-8740-0.ch007

Noguti, V., Singh, S., & Waller, D. S. (2016). Gender Differences in Motivations to Use Social Networking Sites. In R. English & R. Johns (Eds.), *Gender Considerations in Online Consumption Behavior and Internet Use* (pp. 32–49). Hershey, PA: IGI Global. doi:10.4018/978-1-5225-0010-0.ch003

Noor, R. (2017). Citizen Journalism: News Gathering by Amateurs. In M. Adria & Y. Mao (Eds.), *Handbook of Research on Citizen Engagement and Public Participation in the Era of New Media* (pp. 194–229). Hershey, PA: IGI Global. doi:10.4018/978-1-5225-1081-9.ch012

Nwagbara, U., Oruh, E. S., & Brown, C. (2016). State Fragility and Stakeholder Engagement: New Media and Stakeholders' Voice Amplification in the Nigerian Petroleum Industry. In W. Ozuem & G. Bowen (Eds.), *Competitive Social Media Marketing Strategies* (pp. 136–154). Hershey, PA: IGI Global. doi:10.4018/978-1-4666-9776-8.ch007

Obermayer, N., Csepregi, A., & Kővári, E. (2017). Knowledge Sharing Relation to Competence, Emotional Intelligence, and Social Media Regarding Generations. In A. Bencsik (Ed.), *Knowledge Management Initiatives and Strategies in Small and Medium Enterprises* (pp. 269–290). Hershey, PA: IGI Global. doi:10.4018/978-1-5225-1642-2.ch013

Obermayer, N., Gaál, Z., Szabó, L., & Csepregi, A. (2017). Leveraging Knowledge Sharing over Social Media Tools. In R. Chugh (Ed.), *Harnessing Social Media as a Knowledge Management Tool* (pp. 1–24). Hershey, PA: IGI Global. doi:10.4018/978-1-5225-0495-5.ch001

Ogwezzy-Ndisika, A. O., & Faustino, B. A. (2016). Gender Responsive Election Coverage in Nigeria: A Score Card of 2011 General Elections. In L. Mukhongo & J. Macharia (Eds.), *Political Influence of the Media in Developing Countries* (pp. 234–249). Hershey, PA: IGI Global. doi:10.4018/978-1-4666-9613-6.ch015

Okoroafor, O. E. (2018). New Media Technology and Development Journalism in Nigeria. In A. Salawu & T. Owolabi (Eds.), *Exploring Journalism Practice and Perception in Developing Countries* (pp. 105–125). Hershey, PA: IGI Global. doi:10.4018/978-1-5225-3376-4.ch006

Olaleye, S. A., Sanusi, I. T., & Ukpabi, D. C. (2018). Assessment of Mobile Money Enablers in Nigeria. In F. Mtenzi, G. Oreku, D. Lupiana, & J. Yonazi (Eds.), *Mobile Technologies and Socio-Economic Development in Emerging Nations* (pp. 129–155). Hershey, PA: IGI Global. doi:10.4018/978-1-5225-4029-8.ch007

Ozuem, W., Pinho, C. A., & Azemi, Y. (2016). User-Generated Content and Perceived Customer Value. In W. Ozuem & G. Bowen (Eds.), *Competitive Social Media Marketing Strategies* (pp. 50–63). Hershey, PA: IGI Global. doi:10.4018/978-1-4666-9776-8.ch003

Pacchiega, C. (2017). An Informal Methodology for Teaching Through Virtual Worlds: Using Internet Tools and Virtual Worlds in a Coordinated Pattern to Teach Various Subjects. In G. Panconesi & M. Guida (Eds.), *Handbook of Research on Collaborative Teaching Practice in Virtual Learning Environments* (pp. 163–180). Hershey, PA: IGI Global. doi:10.4018/978-1-5225-2426-7.ch009

Pase, A. F., Goss, B. M., & Tietzmann, R. (2018). A Matter of Time: Transmedia Journalism Challenges. In R. Gambarato & G. Alzamora (Eds.), *Exploring Transmedia Journalism in the Digital Age* (pp. 49–66). Hershey, PA: IGI Global. doi:10.4018/978-1-5225-3781-6.ch004

Passarelli, B., & Paletta, F. C. (2016). Living inside the NET: The Primacy of Interactions and Processes. In B. Passarelli, J. Straubhaar, & A. Cuevas-Cerveró (Eds.), *Handbook of Research on Comparative Approaches to the Digital Age Revolution in Europe and the Americas* (pp. 1–15). Hershey, PA: IGI Global. doi:10.4018/978-1-4666-8740-0.ch001

Patkin, T. T. (2017). Social Media and Knowledge Management in a Crisis Context: Barriers and Opportunities. In R. Chugh (Ed.), *Harnessing Social Media as a Knowledge Management Tool* (pp. 125–142). Hershey, PA: IGI Global. doi:10.4018/978-1-5225-0495-5.ch007

Pavlíček, A. (2017). Social Media and Creativity: How to Engage Users and Tourists. In A. Kiráľová (Ed.), *Driving Tourism through Creative Destinations and Activities* (pp. 181–202). Hershey, PA: IGI Global. doi:10.4018/978-1-5225-2016-0.ch009

Pillay, K., & Maharaj, M. (2017). The Business of Advocacy: A Case Study of Greenpeace. In V. Benson, R. Tuninga, & G. Saridakis (Eds.), *Analyzing the Strategic Role of Social Networking in Firm Growth and Productivity* (pp. 59–75). Hershey, PA: IGI Global. doi:10.4018/978-1-5225-0559-4.ch004

Piven, I. P., & Breazeale, M. (2017). Desperately Seeking Customer Engagement: The Five-Sources Model of Brand Value on Social Media. In V. Benson, R. Tuninga, & G. Saridakis (Eds.), *Analyzing the Strategic Role of Social Networking in Firm Growth and Productivity* (pp. 283–313). Hershey, PA: IGI Global. doi:10.4018/978-1-5225-0559-4.ch016

Pokharel, R. (2017). New Media and Technology: How Do They Change the Notions of the Rhetorical Situations? In B. Gurung & M. Limbu (Eds.), *Integration of Cloud Technologies in Digitally Networked Classrooms and Learning Communities* (pp. 120–148). Hershey, PA: IGI Global. doi:10.4018/978-1-5225-1650-7.ch008

Popoola, I. S. (2016). The Press and the Emergent Political Class in Nigeria: Media, Elections, and Democracy. In L. Mukhongo & J. Macharia (Eds.), *Political Influence of the Media in Developing Countries* (pp. 45–58). Hershey, PA: IGI Global. doi:10.4018/978-1-4666-9613-6.ch004

Porlezza, C., Benecchi, E., & Colapinto, C. (2018). The Transmedia Revitalization of Investigative Journalism: Opportunities and Challenges of the Serial Podcast. In R. Gambarato & G. Alzamora (Eds.), *Exploring Transmedia Journalism in the Digital Age* (pp. 183–201). Hershey, PA: IGI Global. doi:10.4018/978-1-5225-3781-6.ch011

Ramluckan, T., Ally, S. E., & van Niekerk, B. (2017). Twitter Use in Student Protests: The Case of South Africa's #FeesMustFall Campaign. In M. Korstanje (Ed.), *Threat Mitigation and Detection of Cyber Warfare and Terrorism Activities* (pp. 220–253). Hershey, PA: IGI Global. doi:10.4018/978-1-5225-1938-6.ch010

Rao, N. R. (2017). Social Media: An Enabler for Governance. In N. Rao (Ed.), *Social Media Listening and Monitoring for Business Applications* (pp. 151–164). Hershey, PA: IGI Global. doi:10.4018/978-1-5225-0846-5.ch008

Rathore, A. K., Tuli, N., & Ilavarasan, P. V. (2016). Pro-Business or Common Citizen?: An Analysis of an Indian Woman CEO's Tweets. *International Journal of Virtual Communities and Social Networking, 8*(1), 19–29. doi:10.4018/IJVCSN.2016010102

Redi, F. (2017). Enhancing Coopetition Among Small Tourism Destinations by Creativity. In A. Kiráľová (Ed.), *Driving Tourism through Creative Destinations and Activities* (pp. 223–244). Hershey, PA: IGI Global. doi:10.4018/978-1-5225-2016-0.ch011

Reeves, M. (2016). Social Media: It Can Play a Positive Role in Education. In R. English & R. Johns (Eds.), *Gender Considerations in Online Consumption Behavior and Internet Use* (pp. 82–95). Hershey, PA: IGI Global. doi:10.4018/978-1-5225-0010-0.ch006

Reis, Z. A. (2016). Bring the Media Literacy of Turkish Pre-Service Teachers to the Table. In M. Yildiz & J. Keengwe (Eds.), *Handbook of Research on Media Literacy in the Digital Age* (pp. 405–422). Hershey, PA: IGI Global. doi:10.4018/978-1-4666-9667-9.ch019

Resuloğlu, F., & Yılmaz, R. (2017). A Model for Interactive Advertising Narration. In R. Yılmaz (Ed.), *Narrative Advertising Models and Conceptualization in the Digital Age* (pp. 1–20). Hershey, PA: IGI Global. doi:10.4018/978-1-5225-2373-4.ch001

Ritzhaupt, A. D., Poling, N., Frey, C., Kang, Y., & Johnson, M. (2016). A Phenomenological Study of Games, Simulations, and Virtual Environments Courses: What Are We Teaching and How? *International Journal of Gaming and Computer-Mediated Simulations, 8*(3), 59–73. doi:10.4018/IJGCMS.2016070104

Ross, D. B., Eleno-Orama, M., & Salah, E. V. (2018). The Aging and Technological Society: Learning Our Way Through the Decades. In V. Bryan, A. Musgrove, & J. Powers (Eds.), *Handbook of Research on Human Development in the Digital Age* (pp. 205–234). Hershey, PA: IGI Global. doi:10.4018/978-1-5225-2838-8.ch010

Rusko, R., & Merenheimo, P. (2017). Co-Creating the Christmas Story: Digitalizing as a Shared Resource for a Shared Brand. In I. Oncioiu (Ed.), *Driving Innovation and Business Success in the Digital Economy* (pp. 137–157). Hershey, PA: IGI Global. doi:10.4018/978-1-5225-1779-5.ch010

Sabao, C., & Chikara, T. O. (2018). Social Media as Alternative Public Sphere for Citizen Participation and Protest in National Politics in Zimbabwe: The Case of #thisflag. In F. Endong (Ed.), *Exploring the Role of Social Media in Transnational Advocacy* (pp. 17–35). Hershey, PA: IGI Global. doi:10.4018/978-1-5225-2854-8.ch002

Samarthya-Howard, A., & Rogers, D. (2018). Scaling Mobile Technologies to Maximize Reach and Impact: Partnering With Mobile Network Operators and Governments. In S. Takavarasha Jr & C. Adams (Eds.), *Affordability Issues Surrounding the Use of ICT for Development and Poverty Reduction* (pp. 193–211). Hershey, PA: IGI Global. doi:10.4018/978-1-5225-3179-1.ch009

Sandoval-Almazan, R. (2017). Political Messaging in Digital Spaces: The Case of Twitter in Mexico's Presidential Campaign. In Y. Ibrahim (Ed.), *Politics, Protest, and Empowerment in Digital Spaces* (pp. 72–90). Hershey, PA: IGI Global. doi:10.4018/978-1-5225-1862-4.ch005

Schultz, C. D., & Dellnitz, A. (2018). Attribution Modeling in Online Advertising. In K. Yang (Ed.), *Multi-Platform Advertising Strategies in the Global Marketplace* (pp. 226–249). Hershey, PA: IGI Global. doi:10.4018/978-1-5225-3114-2.ch009

Schultz, C. D., & Holsing, C. (2018). Differences Across Device Usage in Search Engine Advertising. In K. Yang (Ed.), *Multi-Platform Advertising Strategies in the Global Marketplace* (pp. 250–279). Hershey, PA: IGI Global. doi:10.4018/978-1-5225-3114-2.ch010

Senadheera, V., Warren, M., Leitch, S., & Pye, G. (2017). Facebook Content Analysis: A Study into Australian Banks' Social Media Community Engagement. In S. Hai-Jew (Ed.), *Social Media Data Extraction and Content Analysis* (pp. 412–432). Hershey, PA: IGI Global. doi:10.4018/978-1-5225-0648-5.ch013

Sharma, A. R. (2018). Promoting Global Competencies in India: Media and Information Literacy as Stepping Stone. In M. Yildiz, S. Funk, & B. De Abreu (Eds.), *Promoting Global Competencies Through Media Literacy* (pp. 160–174). Hershey, PA: IGI Global. doi:10.4018/978-1-5225-3082-4.ch010

Sillah, A. (2017). Nonprofit Organizations and Social Media Use: An Analysis of Nonprofit Organizations' Effective Use of Social Media Tools. In M. Brown Sr., (Ed.), *Social Media Performance Evaluation and Success Measurements* (pp. 180–195). Hershey, PA: IGI Global. doi:10.4018/978-1-5225-1963-8.ch009

Škorić, M. (2017). Adaptation of Winlink 2000 Emergency Amateur Radio Email Network to a VHF Packet Radio Infrastructure. In A. El Oualkadi & J. Zbitou (Eds.), *Handbook of Research on Advanced Trends in Microwave and Communication Engineering* (pp. 498–528). Hershey, PA: IGI Global. doi:10.4018/978-1-5225-0773-4.ch016

Skubida, D. (2016). Can Some Computer Games Be a Sport?: Issues with Legitimization of eSport as a Sporting Activity. *International Journal of Gaming and Computer-Mediated Simulations*, *8*(4), 38–52. doi:10.4018/IJGCMS.2016100103

Sonnenberg, C. (2016). Mobile Content Adaptation: An Analysis of Techniques and Frameworks. In J. Aguado, C. Feijóo, & I. Martínez (Eds.), *Emerging Perspectives on the Mobile Content Evolution* (pp. 177–199). Hershey, PA: IGI Global. doi:10.4018/978-1-4666-8838-4.ch010

Sonnevend, J. (2016). More Hope!: Ceremonial Media Events Are Still Powerful in the Twenty-First Century. In A. Fox (Ed.), *Global Perspectives on Media Events in Contemporary Society* (pp. 132–140). Hershey, PA: IGI Global. doi:10.4018/978-1-4666-9967-0.ch010

Sood, T. (2017). Services Marketing: A Sector of the Current Millennium. In T. Sood (Ed.), *Strategic Marketing Management and Tactics in the Service Industry* (pp. 15–42). Hershey, PA: IGI Global. doi:10.4018/978-1-5225-2475-5.ch002

Stairs, G. A. (2016). The Amplification of the Sunni-Shia Divide through Contemporary Communications Technology: Fear and Loathing in the Modern Middle East. In S. Gibson & A. Lando (Eds.), *Impact of Communication and the Media on Ethnic Conflict* (pp. 214–231). Hershey, PA: IGI Global. doi:10.4018/978-1-4666-9728-7.ch013

Stokinger, E., & Ozuem, W. (2016). The Intersection of Social Media and Customer Retention in the Luxury Beauty Industry. In W. Ozuem & G. Bowen (Eds.), *Competitive Social Media Marketing Strategies* (pp. 235–258). Hershey, PA: IGI Global. doi:10.4018/978-1-4666-9776-8.ch012

Sudarsanam, S. K. (2017). Social Media Metrics. In N. Rao (Ed.), *Social Media. Listening and Monitoring for Business Applications* (pp. 131–149). Hershey, PA: IGI Global. doi:10.4018/978-1-5225-0846-5.ch007

Swiatek, L. (2017). Accessing the Finest Minds: Insights into Creativity from Esteemed Media Professionals. In N. Silton (Ed.), *Exploring the Benefits of Creativity in Education, Media, and the Arts* (pp. 240–263). Hershey, PA: IGI Global. doi:10.4018/978-1-5225-0504-4.ch012

Switzer, J. S., & Switzer, R. V. (2016). Virtual Teams: Profiles of Successful Leaders. In B. Baggio (Ed.), *Analyzing Digital Discourse and Human Behavior in Modern Virtual Environments* (pp. 1–24). Hershey, PA: IGI Global. doi:10.4018/978-1-4666-9899-4.ch001

Tabbane, R. S., & Debabi, M. (2016). Electronic Word of Mouth: Definitions and Concepts. In S. Rathore & A. Panwar (Eds.), *Capturing, Analyzing, and Managing Word-of-Mouth in the Digital Marketplace* (pp. 1–27). Hershey, PA: IGI Global. doi:10.4018/978-1-4666-9449-1.ch001

Tellería, A. S. (2016). The Role of the Profile and the Digital Identity on the Mobile Content. In J. Aguado, C. Feijóo, & I. Martínez (Eds.), *Emerging Perspectives on the Mobile Content Evolution* (pp. 263–282). Hershey, PA: IGI Global. doi:10.4018/978-1-4666-8838-4.ch014

Teurlings, J. (2017). What Critical Media Studies Should Not Take from Actor-Network Theory. In M. Spöhrer & B. Ochsner (Eds.), *Applying the Actor-Network Theory in Media Studies* (pp. 66–78). Hershey, PA: IGI Global. doi:10.4018/978-1-5225-0616-4.ch005

Tomé, V. (2018). Assessing Media Literacy in Teacher Education. In M. Yildiz, S. Funk, & B. De Abreu (Eds.), *Promoting Global Competencies Through Media Literacy* (pp. 1–19). Hershey, PA: IGI Global. doi:10.4018/978-1-5225-3082-4.ch001

Toscano, J. P. (2017). Social Media and Public Participation: Opportunities, Barriers, and a New Framework. In M. Adria & Y. Mao (Eds.), *Handbook of Research on Citizen Engagement and Public Participation in the Era of New Media* (pp. 73–89). Hershey, PA: IGI Global. doi:10.4018/978-1-5225-1081-9.ch005

Trauth, E. (2017). Creating Meaning for Millennials: Bakhtin, Rosenblatt, and the Use of Social Media in the Composition Classroom. In K. Bryant (Ed.), *Engaging 21st Century Writers with Social Media* (pp. 151–162). Hershey, PA: IGI Global. doi:10.4018/978-1-5225-0562-4.ch009

Ugangu, W. (2016). Kenya's Difficult Political Transitions Ethnicity and the Role of Media. In L. Mukhongo & J. Macharia (Eds.), *Political Influence of the Media in Developing Countries* (pp. 12–24). Hershey, PA: IGI Global. doi:10.4018/978-1-4666-9613-6.ch002

Uprety, S. (2018). Print Media's Role in Securitization: National Security and Diplomacy Discourses in Nepal. In S. Bute (Ed.), *Media Diplomacy and Its Evolving Role in the Current Geopolitical Climate* (pp. 56–82). Hershey, PA: IGI Global. doi:10.4018/978-1-5225-3859-2.ch004

Van der Merwe, L. (2016). Social Media Use within Project Teams: Practical Application of Social Media on Projects. In G. Silvius (Ed.), *Strategic Integration of Social Media into Project Management Practice* (pp. 139–159). Hershey, PA: IGI Global. doi:10.4018/978-1-4666-9867-3.ch009

van der Vyver, A. G. (2018). A Model for Economic Development With Telecentres and the Social Media: Overcoming Affordability Constraints. In S. Takavarasha Jr & C. Adams (Eds.), *Affordability Issues Surrounding the Use of ICT for Development and Poverty Reduction* (pp. 112–140). Hershey, PA: IGI Global. doi:10.4018/978-1-5225-3179-1.ch006

van Dokkum, E., & Ravesteijn, P. (2016). Managing Project Communication: Using Social Media for Communication in Projects. In G. Silvius (Ed.), *Strategic Integration of Social Media into Project Management Practice* (pp. 35–50). Hershey, PA: IGI Global. doi:10.4018/978-1-4666-9867-3.ch003

van Niekerk, B. (2018). Social Media Activism From an Information Warfare and Security Perspective. In F. Endong (Ed.), *Exploring the Role of Social Media in Transnational Advocacy* (pp. 1–16). Hershey, PA: IGI Global. doi:10.4018/978-1-5225-2854-8.ch001

Varnali, K., & Gorgulu, V. (2017). Determinants of Brand Recall in Social Networking Sites. In W. Al-Rabayah, R. Khasawneh, R. Abu-shamaa, & I. Alsmadi (Eds.), *Strategic Uses of Social Media for Improved Customer Retention* (pp. 124–153). Hershey, PA: IGI Global. doi:10.4018/978-1-5225-1686-6.ch007

Varty, C. T., O'Neill, T. A., & Hambley, L. A. (2017). Leading Anywhere Workers: A Scientific and Practical Framework. In Y. Blount & M. Gloet (Eds.), *Anywhere Working and the New Era of Telecommuting* (pp. 47–88). Hershey, PA: IGI Global. doi:10.4018/978-1-5225-2328-4.ch003

Vatikiotis, P. (2016). Social Media Activism: A Contested Field. In T. Deželan & I. Vobič (Eds.), *R)evolutionizing Political Communication through Social Media* (pp. 40–54). Hershey, PA: IGI Global. doi:10.4018/978-1-4666-9879-6.ch003

Velikovsky, J. T. (2018). The Holon/Parton Structure of the Meme, or The Unit of Culture. In M. Khosrow-Pour, D.B.A. (Ed.), Encyclopedia of Information Science and Technology, Fourth Edition (pp. 4666-4678). Hershey, PA: IGI Global. doi:10.4018/978-1-5225-2255-3.ch405

Venkatesh, R., & Jayasingh, S. (2017). Transformation of Business through Social Media. In N. Rao (Ed.), *Social Media Listening and Monitoring for Business Applications* (pp. 1–17). Hershey, PA: IGI Global. doi:10.4018/978-1-5225-0846-5.ch001

Vesnic-Alujevic, L. (2016). European Elections and Facebook: Political Advertising and Deliberation? In T. Deželan & I. Vobič (Eds.), *R)evolutionizing Political Communication through Social Media* (pp. 191–209). Hershey, PA: IGI Global. doi:10.4018/978-1-4666-9879-6.ch010

Virkar, S. (2017). Trolls Just Want to Have Fun: Electronic Aggression within the Context of E-Participation and Other Online Political Behaviour in the United Kingdom. In M. Korstanje (Ed.), *Threat Mitigation and Detection of Cyber Warfare and Terrorism Activities* (pp. 111–162). Hershey, PA: IGI Global. doi:10.4018/978-1-5225-1938-6.ch006

Wakabi, W. (2017). When Citizens in Authoritarian States Use Facebook for Social Ties but Not Political Participation. In Y. Ibrahim (Ed.), *Politics, Protest, and Empowerment in Digital Spaces* (pp. 192–214). Hershey, PA: IGI Global. doi:10.4018/978-1-5225-1862-4.ch012

Weisberg, D. J. (2016). Methods and Strategies in Using Digital Literacy in Media and the Arts. In M. Yildiz & J. Keengwe (Eds.), *Handbook of Research on Media Literacy in the Digital Age* (pp. 456–471). Hershey, PA: IGI Global. doi:10.4018/978-1-4666-9667-9.ch022

Weisgerber, C., & Butler, S. H. (2016). Debranding Digital Identity: Personal Branding and Identity Work in a Networked Age. *International Journal of Interactive Communication Systems and Technologies*, 6(1), 17–34. doi:10.4018/IJICST.2016010102

Wijngaard, P., Wensveen, I., Basten, A., & de Vries, T. (2016). Projects without Email, Is that Possible? In G. Silvius (Ed.), *Strategic Integration of Social Media into Project Management Practice* (pp. 218–235). Hershey, PA: IGI Global. doi:10.4018/978-1-4666-9867-3.ch013

Wright, K. (2018). "Show Me What You Are Saying": Visual Literacy in the Composition Classroom. In A. August (Ed.), *Visual Imagery, Metadata, and Multimodal Literacies Across the Curriculum* (pp. 24–49). Hershey, PA: IGI Global. doi:10.4018/978-1-5225-2808-1.ch002

Yang, K. C. (2018). Understanding How Mexican and U.S. Consumers Decide to Use Mobile Social Media: A Cross-National Qualitative Study. In K. Yang (Ed.), *Multi-Platform Advertising Strategies in the Global Marketplace* (pp. 168–198). Hershey, PA: IGI Global. doi:10.4018/978-1-5225-3114-2.ch007

Yang, K. C., & Kang, Y. (2016). Exploring Female Hispanic Consumers' Adoption of Mobile Social Media in the U.S. In R. English & R. Johns (Eds.), *Gender Considerations in Online Consumption Behavior and Internet Use* (pp. 185–207). Hershey, PA: IGI Global. doi:10.4018/978-1-5225-0010-0.ch012

Yao, Q., & Wu, M. (2016). Examining the Role of WeChat in Advertising. In X. Xu (Ed.), *Handbook of Research on Human Social Interaction in the Age of Mobile Devices* (pp. 386–405). Hershey, PA: IGI Global. doi:10.4018/978-1-5225-0469-6.ch016

Yarchi, M., Wolfsfeld, G., Samuel-Azran, T., & Segev, E. (2017). Invest, Engage, and Win: Online Campaigns and Their Outcomes in an Israeli Election. In M. Brown Sr., (Ed.), *Social Media Performance Evaluation and Success Measurements* (pp. 225–248). Hershey, PA: IGI Global. doi:10.4018/978-1-5225-1963-8.ch011

Yeboah-Banin, A. A., & Amoakohene, M. I. (2018). The Dark Side of Multi-Platform Advertising in an Emerging Economy Context. In K. Yang (Ed.), *Multi-Platform Advertising Strategies in the Global Marketplace* (pp. 30–53). Hershey, PA: IGI Global. doi:10.4018/978-1-5225-3114-2.ch002

Yılmaz, R., Çakır, A., & Resuloğlu, F. (2017). Historical Transformation of the Advertising Narration in Turkey: From Stereotype to Digital Media. In R. Yılmaz (Ed.), *Narrative Advertising Models and Conceptualization in the Digital Age* (pp. 133–152). Hershey, PA: IGI Global. doi:10.4018/978-1-5225-2373-4.ch008

Yusuf, S., Hassan, M. S., & Ibrahim, A. M. (2018). Cyberbullying Among Malaysian Children Based on Research Evidence. In M. Khosrow-Pour, D.B.A. (Ed.), Encyclopedia of Information Science and Technology, Fourth Edition (pp. 1704-1722). Hershey, PA: IGI Global. doi:10.4018/978-1-5225-2255-3.ch149

Zervas, P., & Alexandraki, C. (2016). Facilitating Open Source Software and Standards to Assembly a Platform for Networked Music Performance. In D. Kanellopoulos (Ed.), *Emerging Research on Networked Multimedia Communication Systems* (pp. 334–365). Hershey, PA: IGI Global. doi:10.4018/978-1-4666-8850-6.ch011

About the Contributors

Jyotsana Thakur is working as an Assistant Professor in Media Studies. She has a M. Phil and PhD in Journalism & Mass Communication and teaches both UG and PG Classes. She has 6 years of industry experience as Public Relations Officer and 8 years of teaching experience. Her areas of interest are Advertising, Corporate Communication, Brand Management, Social Media, New Media, Development and Communication, Small and Large Businesses, ICT, E-Learning, Blended Learning, Media Laws & Ethics, Consumer Satisfaction, Corporate Reputation, Issue Management and Crisis Management, Corporate Social Responsibilities, E-Commerce and E-Governance, Communication Strategies, Event Management. She is a Mentor and involved in various committees in her organization. She has published many research papers in reputed journals and presented/attended papers in both national and international conferences, seminars and workshops. Got an award "BUDDING RESEARCHER OF THE YEAR 2016" by the International Association of Research & Development Organization. She is a member of advisory/reviewer board of three reputed journals. She is a member of editorial boards in many National and International Journals of high repute.

* * *

Azza A. Ahmed is a professor of Mass Communication at Abu Dhabi University, and the Faculty of Mass Communication, Cairo University in Egypt. She has publications in regional and international refereed journals published in USA, Canada and Egypt. Prof. Azza participated in many international refereed conferences in UK, USA, Bahrain, Singapore, Malaysia, China and Spain. She got many international research awards from Virginia, Georgia State Universities in USA. She is an active member is many International Media associations such as the Arab-US Association for Media Educators.

Mahmoud Bakkar is an innovative person, with an extensive experience in teaching and training in UAE and Australia. He also has an expansive knowledge of business development, management and leadership. He is currently working as a lecturer at Holmes Institute in Melbourne. Mahmoud has previously held academic positions as a lecturer in MIT university in Melbourne and IT instructor in Higher College of Technology, in UAE-Dubai. He has also, held IT industry positions in Philips Middle East and worked as SIEMENS Customer Support Engineer in UAE- Dubai. Due to his passion with item response theory, Mahmoud recently initiated the Rasch Model Group (RMG) in the social networking site Research-Gate. Mahmoud's research interests involve: IT/IS capabilities, Data Mining; Big Data; Business Analytics; Cognitive Computing and their application in Business; Health and Education; eBusiness; Digital Business; Intelligent Systems; Decision Analysis; Optimization; Evolutionary Computing; Human-computer Interaction; Digital Business; Information Systems Development & Design; IT Innovation & Management; Cloud Computing; Cyber Security, IT in Education, Information Systems Security; Cloud Computing and Mobile Computing; new developments in ICT; ICT for Education; Knowledge Measurement; Assessment Measurement; Social Networks and Media.

Allaa Barefah is a goals-oriented PhD scholar (in her final submission year) in the School of Information Technology and Logistics, at RMIT University, Melbourne, Australia. Allaa's research focuses on investigating effective ePedagogical practices that are employed in eLearning courseware for the higher education sector. She is interested in improving the teaching and learning practice of information systems domain-specific courses. Allaa holds a Master Degree (with distinction) in Business Information Technology, from RMIT University, and has teaching experience in information systems analysis and design, database management, and information technology networking. Her works involve the utilisation of experimental designs and Rasch evaluation models.

Sarah Bouraga has a PhD from and is a lecturer at the University of Namur, Department of Business Administration. She is a member of the PReCISE Research Center. Her main research interests include Requirements Engineering, Online Social Networks, Decision-Making, and Knowledge Engineering. More specifically, she is interested in how to conduct Requirements engineering for the design of a new online social network and its knowledge-based recommendation system. Her other research interests include Business Intelligence, Big Data and Data Valuation.

P. Vigneswara Ilavarasan (PhD – IIT Kanpur) is Associate Professor at the Dept. of Management Studies, Indian Institute of Technology Delhi. He researches and teaches about production and consumption of information and communication technologies (ICTs) with a special focus on India. His publications and other details are available at: http://web.iitd.ac.in/~vignes/.

Ivan Jureta is Tenured Researcher with the National Fund for Scientific Research in Belgium, and Honorary Associate Professor at Department of Business Administration and Department of Computer Science at University of Namur. Ivan is the author of two books and over 80 research papers on knowledge engineering for problem solving and decision making. He is involved with several startups at CxO level and has held lead roles in product design for web and digital services that today serve more than 500.000 people every day.

Elspeth McKay, PhD, Fellow ACS, is located in the School of Business IT and Logistics at the RMITI University, Australia. Her PhD is in Computer Science and Information Systems, from Deakin University, Geelong, Australia. Elspeth also holds further qualifications in Instructional Design, Computer Education and Business Information Systems. She is passionate about designing effective eLearning resources for the education sector and industry training/reskilling programmes. Her Australian Research Council's research project investigated Government eTraining strategies. Her work involves developing specialist e-Learning tools implemented through rich internet applications; including: ARPS – an advanced repurposing pilot system, COGNIWARE – a multi-modal e-Learning framework, GEMS – a global eMuseum System, eWRAP – Electronic work readiness awareness programme, EASY – Educational/academic (skills) screening for the young, offering enhanced accessibility through touch screen technologies. Over the last decade Dr McKay has published extensively in the research fields of HCI and educational technology.

Marlina Mohamad is a senior lecturer at Faculty of Technical and Vocational Education, Universiti Tun Hussein Onn Malaysia. Her PhD is in Business Information Systems (Instructional Design) from RMIT University, Melbourne, Australia. She also holds a Master Degree in Information Technology (Management) and a Degree in Computer Science (Software Engineering). Her research interest includes: web-mediated learning and training; e-pedagogy; instructional technology and design; ICT in Technical Vocational Education; and Training and measurement.

Ashish Rathore is a PhD Scholar in the Department of Management Studies, Indian Institute of Technology Delhi. He holds Maters and Bachelor Degree in Mechanical Engineering. His current research interests include Social Media, Product development and Data Mining.

Deepali Saluja is Assistant Professor in the Banarsidas Chandiwala Institute of Professional Studies, New Delhi, India. She has been awarded by Ph. D. in the area of Economics. She is also NET qualified in Economics. She has experience in academics more than thirteen years in which she has supervised many projects of students at postgraduate and undergraduate level. She presented papers in National and International conference and attended many conferences, workshops and FDPs. and N.S.S. She also won the President and Governor Award in Scouting Guiding. She has authored several paper published in national and International refereed journals.

Shamsher Singh is Associate Professor in the Banarsidas Chandiwala Institute of Professional Studies, New Delhi, India. He received his PhD in Management from Jamia Hamdard University, New Delhi (Joint Programme of Jamia Hamdard University, New Delhi and MDI Gurgaon). His topic of research was 'The Study of Effectiveness of Customer Relationship Management in Indian Banking Sector with special reference to NCR Delhi'. He completed his MBA degree with specialisation in Marketing Management from Department of Management Science, University of Pune (PUMBA) in 1996, Post Graduate Diploma in Export-Import Management from Indian Merchant Chamber, Mumbai in 1998, Post Graduate Diploma in Human Resources Management from IGNOU, in 2002, and Post Graduate Degree in Public Administration from Panjab University, in 1992.He has authored several papers published in national and International refereed journals and has also presented paper in national and international conferences . He has also authored three text books and edited two books.

Petter Skult is currently writing as a Game Designer at Snowfall, a game development company in Finland. His doctoral dissertation focuses on post-apocalyptic science fiction, with a emphasis on the novels of Paul Auster, Cormac McCarthy and Margaret Atwood. His research interests include the burgeoning pop-cultural fascination with everything apocalyptic/post-apocalyptic in the post-Cold War era, including zombie fiction. His research also includes the evolution of literary theory from postmodernism to what might be termed post-postmodernism, and the connection between science fiction and post-postmodern literature. More information can be found at www.abo.fi/fakultet/Content/Document/document/27269 and www.linkedin.com/in/petter-skult-6a7b37b2/.

Jouni Smed holds a PhD in Computer Science and acts as Senior Lecturer and Adjunct Professor in the University of Turku, Finland. He has organized and taught game development on diverse topics ranging from game algorithms and networking in multiplayer games to game software construction, game design and interactive storytelling. He is also the co-founder of Turku Game Lab, which aims at bringing together technologically- and artistically-oriented students to work on the same game projects and start their career in the game industry. His research interests range from code tweaking to software processes and from simple puzzles to multisite game development.

Tomi "bgt" Suovuo is a Ph.D. student at University of Turku, Department of Future Technologies, Subject of Interaction Design. He is teaching and researching computer mediated interaction, particularly games and interactive storytelling.

Natasha Trygg is the CEO of Snowfall, a leading developer of Moomin mobile games. She is an active member of the Finnish gaming industry scene as board member of Neogames, IGDA Finland and Chairperson of The Hive - Turku Game Hub. Her expertise and established position in the industry along with her academic work as Doctoral Candidate and teacher at University of Turku, gives her super-powers in her multidisciplinary approach to every challenge that presents itself. In her free time, she is a passionate gamer who mainly enjoys playing PC and board games, although VR experiences are every day becoming a bigger part of her retirement plans. More information can be found at www.natashatrygg.com and www.linkedin.com/in/natasha-bulatovic-trygg.

Nikhil Tuli graduated from Department of Management Studies of IIT Delhi.

Index

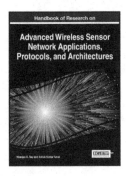

Ensure Quality Research is Introduced to the Academic Community

Become an IGI Global Reviewer for Authored Book Projects

Premier Reference Source

Emerging GIS Applications for Emergency and Disaster Management

Premier Reference Source

Managerial Strategies and Green Solutions for Project Sustainability

Premier Reference Source

Comparative Approaches to Using R and Python for Statistical Data Analysis

Premier Reference Source

Solutions for High-Touch Communications in a High-Tech World

The overall success of an authored book project is dependent on quality and timely reviews.

In this competitive age of scholarly publishing, constructive and timely feedback significantly expedites the turnaround time of manuscripts from submission to acceptance, allowing the publication and discovery of forward-thinking research at a much more expeditious rate. Several IGI Global authored book projects are currently seeking highly qualified experts in the field to fill vacancies on their respective editorial review boards:

Applications may be sent to:
development@igi-global.com

Applicants must have a doctorate (or an equivalent degree) as well as publishing and reviewing experience. Reviewers are asked to write reviews in a timely, collegial, and constructive manner. All reviewers will begin their role on an ad-hoc basis for a period of one year, and upon successful completion of this term can be considered for full editorial review board status, with the potential for a subsequent promotion to Associate Editor.

If you have a colleague that may be interested in this opportunity, we encourage you to share this information with them.

Printed in the United States
By Bookmasters